Persistent Organic Pollutants

Persistent Organic Pollutants

Edited by
Kaleb Lynch

Larsen & Keller
www.larsen-keller.com

Persistent Organic Pollutants
Edited by Kaleb Lynch
ISBN: 978-1-63549-211-8 (Hardback)

© 2017 Larsen & Keller

▤ Larsen & Keller

Published by Larsen and Keller Education,
5 Penn Plaza,
19th Floor,
New York, NY 10001, USA

Cataloging-in-Publication Data

Persistent organic pollutants / edited by Kaleb Lynch.
    p. cm.
Includes bibliographical references and index.
ISBN 978-1-63549-211-8
1. Organic compounds--Toxicology. 2. Organic compounds--Environmental aspects.
3. Persistent pollutants--Environmental aspects. I. Lynch, Kaleb.
RA1235 .P47 2017
615.95--dc23

The publisher's policy is to use permanent paper from mills that operate a sustainable forestry policy. Furthermore, the publisher ensures that the text paper and cover boards used have met acceptable environmental accreditation standards.

Printed and bound in the United States of America.

For more information regarding Larsen and Keller Education and its products, please visit the publisher's website www.larsen-keller.com

Table of Contents

Preface

Organic pollutants, also referred to as persistent organic pollutants are those compounds which are resistant to every type of environmental degradation by natural processes. Such non-biodegradable compounds cause considerable harm and also prove to be highly difficult to remove. This book presents the complex subject of organic pollutants in the most comprehensible and easy to understand language. The topics covered in this extensive book deal with the core subjects of this area. It unfolds the innovative aspects of this discipline which will be crucial for the holistic understanding of the subject matter. Coherent flow of topics, student-friendly language and extensive use of examples make this book an invaluable source of knowledge.

To facilitate a deeper understanding of the contents of this book a short introduction of every chapter is written below:

Chapter 1- Persistent organic pollutants are organic compounds that are unaffected to environmental degradation. They are also used as pesticides and solvents. Persistent organic pollutants are either man-made or can arise naturally with naturally persistent occurring organic pollutants being volcanoes and various biosynthetic pathways. This chapter will provide an integrated understanding of persistent organic pollutants.

Chapter 2- The processes concerned within this text are chemical decomposition, biodegradation, photodissociation and photocatalysis. Chemical decomposition is the breakdown of chemical compounds into simpler compounds whereas the fragmentation of materials by bacteria or fungi is known as biodegradation. The major processes of toxic degradation are dealt with great details in the text. The aspects elucidated in this chapter are of vital importance, and provide a better understanding of persistent organic pollutants.

Chapter 3- The compounds that generate organic pollutants are organobromines, flame-retardants, organochlorides and dioxins and dioxin-like compounds. Organochlorides and organochlorine compounds contain at least one atom of chlorine. Flame retardants are the chemical substances that are added to either plastic or textiles. The aspects elucidated in this chapter are of vital importance, and provide a better understanding of organobromine and organochloride compounds.

Chapter 4- Pollutants that contaminate agricultural practices can be classified into insects, DDT, fungicides and pesticides. Insecticides are substances used in extermination of insects and fungicides are used to kill fungi or fungal spores. The following section is a compilation of the various branches of agricultural pollutants that form an integral part of the broader subject matter.

Chapter 5- The text explains organic pollutants and their effects on our environment. Some of the topics covered are bioaccumulation, biomagnification, bioconcentration and environmental impacts of pesticides. The collection of pesticides and other chemicals in an organism is known as bioaccumulation whereas the effects that pesticides leave on species is discussed in the segment related to environmental impact of pesticides. The following chapter explains the importance of understanding organic pollutants and their effects on the environment.

Chapter 6- Organic pollutants have adverse effects on humans as well as our environment. Endocrine disruptor and neurotoxins are toxins of organic pollutants; endocrine disruptors cause tumors and birth defects whereas neurotoxins are known for causing nervous system related problems. This chapter helps the reader in developing an in-depth understanding of the impact that organic pollutants have on humans.

I owe the completion of this book to the never-ending support of my family, who supported me throughout the project.

Editor

Introduction to Organic Pollutants

Persistent organic pollutants are organic compounds that are unaffected to environmental degradation. They are also used as pesticides and solvents. Persistent organic pollutants are either man-made or can arise naturally with naturally persistent occurring organic pollutants being volcanoes and various biosynthetic pathways. This chapter will provide an integrated understanding of persistent organic pollutants.

Persistent organic pollutants (POPs) are organic compounds that are resistant to environmental degradation through chemical, biological, and photolytic processes. Because of their persistence, POPs bioaccumulate with potential significant impacts on human health and the environment. The effect of POPs on human and environmental health was discussed, with intention to eliminate or severely restrict their production, by the international community at the Stockholm Convention on Persistent Organic Pollutants in 2001.

Many POPs are currently or were in the past used as pesticides, solvents, pharmaceuticals, and industrial chemicals. Although some POPs arise naturally, for example volcanoes and various biosynthetic pathways, most are man-made via total synthesis.

Consequences of Persistence

POPs typically are halogenated organic compounds and as such exhibit high lipid solubility. For this reason, they bioaccumulate in fatty tissues. Halogenated compounds also exhibit great stability reflecting the nonreactivity of C-Cl bonds toward hydrolysis and photolytic degradation. The stability and lipophilicity of organic compounds often correlates with their halogen content, thus polyhalogenated organic compounds are of particular concern. They exert their negative effects on the environment through two processes, long range transport, which allows them to travel far from their source, and bioaccumulation, which reconcentrates these chemical compounds to potentially dangerous levels. Compounds that make up POPs are also classed as PBTs (Persistent, Bioaccumulative and Toxic) or TOMPs (Toxic Organic Micro Pollutants).

Long-range Transport

POPs enter the gas phase under certain environmental temperatures and volatize from soils, vegetation, and bodies of water into the atmosphere, resisting breakdown reactions in the air, to travel long distances before being re-deposited. This results in accumulation of POPs in areas far from where they were used or emitted, specifically environments where POPs have never been introduced such as Antarctica, and the Arctic circle. POPs can be present as vapors in the atmosphere or bound to the surface of solid particles. POPs have low solubility in water but are easily captured by solid particles, and are soluble in organic fluids (oils, fats, and liquid fuels). POPs are not easily degraded in the environment due to their stability and low decomposition rates. Due to this capacity for long-range transport, POP environmental contamination is extensive, even in areas where POPs have never been used, and will remain in these environments years after restrictions implemented due to their resistance to degradation.

Bioaccumulation

Bioaccumulation of POPs is typically associated with the compounds high lipid solubility and ability to accumulate in the fatty tissues of living organisms for long periods of time. Persistent chemicals tend to have higher concentrations and are eliminated more slowly. Dietary accumulation or bioaccumulation is another hallmark characteristic of POPs, as POPs move up the food chain, they increase in concentration as they are processed and metabolized in certain tissues of organisms. The natural capacity for animals gastrointestinal tract concentrate ingested chemicals, along with poorly metabolized and hydrophobic nature of POPs makes such compounds highly susceptible to bioaccumulation. Thus POPs not only persist in the environment, but also as they are taken in by animals they bioaccumulate, increasing their concentration and toxicity in the environment.

Stockholm Convention on Persistent Organic Pollutants

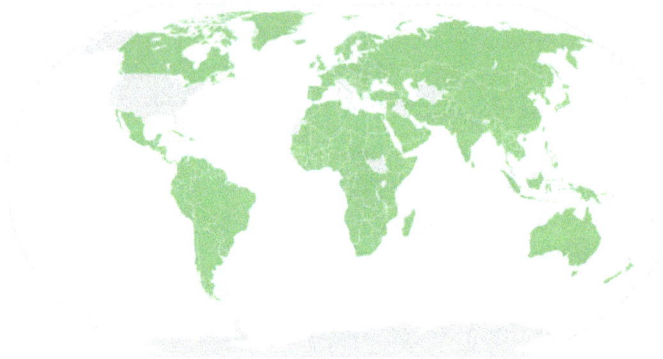

State parties to the Stockholm Convention on Persistent Organic Pollutants

The Stockholm Convention was adopted and put into practice by the United Nations Environment Programme (UNEP) on May 22, 2001. The UNEP decided that POP regulation needed to be addressed globally for the future. The purpose statement of the agreement is "to protect human health and the environment from persistent organic pollutants." As of 2014, there are 179 countries in compliance with the Stockholm convention. The convention and its participants have recognized the potential human and environmental toxicity of POPs. They recognize that POPs have the potential for long range transport and bioaccumulation and biomagnification. The convention seeks to study and then judge whether or not a number of chemicals that have been developed with advances in technology and science can be categorized as POPs or not. The initial meeting in 2001 made a preliminary list, termed the "dirty dozen," of chemicals that are classified as POPs. As of 2014, the United States of America has signed the Stockholm Convention but has not ratified it. There are a handful of other countries that have not ratified the convention but most countries in the world have ratified the convention.

Compounds on The Stockholm Convention List

In May 1995, the United Nations Environment Programme Governing Council investigated POPs. Initially the Convention recognized only twelve POPs for their adverse effects on human health and the environment, placing a global ban on these particularly harmful and toxic compounds and requiring its parties to take measures to eliminate or reduce the release of POPs in the environment.

1. Aldrin, an insecticide used in soils to kill termites, grasshoppers, Western corn rootworm, and others, is also known to kill birds, fish, and humans. Humans are primarily exposed to aldrin through dairy products and animal meats.

2. Chlordane, an insecticide used to control termites and on a range of agricultural crops, is known to be lethal in various species of birds, including mallard ducks, bobwhite quail, and pink shrimp; it is a chemical that remains in the soil with a reported half-life of one year. Chlordane has been postulated to affect the human immune system and is classified as a possible human carcinogen. Chlordane air pollution is believed the primary route of humane exposure.

3. Dieldrin, a pesticide used to control termites, textile pests, insect-borne diseases and insects living in agricultural soils. In soil and insects, aldrin can be oxidized, resulting in rapid conversion to dieldrin. Dieldrin's half-life is approximately five years. Dieldrin is highly toxic to fish and other aquatic animals, particularly frogs, whose embryos can develop spinal deformities after exposure to low levels. Dieldrin has been linked to Parkinson's disease, breast cancer, and classified as immunotoxic, neurotoxic, with endocrine disrupting capacity. Dieldrin residues have been found in air, water, soil, fish, birds, and mammals. Human exposure to dieldrin primarily derives from food.

4. Endrin, an insecticide sprayed on the leaves of crops, and used to control rodents. Animals can metabolize endrin, so fatty tissue accumulation is not an issue, however the chemical has a long half-life in soil for up to 12 years. Endrin is highly toxic to aquatic animals and humans as a neurotoxin. Human exposure results primarily through food.

5. Heptachlor, a pesticide primarily used to kill soil insects and termites, along with cotton insects, grasshoppers, other crop pests, and malaria-carrying mosquitoes. Heptachlor, even at every low doses has been associated with the de-

cline of several wild bird populations – Canada geese and American kestrels. In laboratory tests have shown high-dose heptachlor as lethal, with adverse behavioral changes and reduced reproductive success at low-doses, and is classified as a possible human carcinogen. Human exposure primarily results from food.

6. Hexachlorobenzene (HCB), was first introduced in 1945–1959 to treat seeds because it can kill fungi on food crops. HCB-treated seed grain consumption is associated with photosensitive skin lesions, colic, debilitation, and a metabolic disorder called porphyria turcica, which can be lethal. Mothers who pass HCB to their infants through the placenta and breast milk had limited reproductive success including infant death. Human exposure is primarily from food.

7. Mirex, an insecticide used against ants and termites or as a flame retardant in plastics, rubber, and electrical goods. Mirex is one of the most stable and persistent pesticides, with a half-life of up to 10 years. Mirex is toxic to several plant, fish and crustacean species, with suggested carcinogenic capacity in humans. Humans are exposed primarily through animal meat, fish, and wild game.

8. Toxaphene, an insecticide used on cotton, cereal, grain, fruits, nuts, and vegetables, as well as for tick and mite control in livestock. Widespread toxaphene use in the US and chemical persistence, with a half-life of up to 12 years in soil, results in residual toxaphene in the environment. Toxaphene is highly toxic to fish, inducing dramatic weight loss and reduced egg viability. Human exposure primarily results from food. While human toxicity to direct toxaphene exposure is low, the compound is classified as a possible human carcinogen.

9. Polychlorinated biphenyls (PCBs), used as heat exchange fluids, in electrical transformers, and capacitors, and as additives in paint, carbonless copy paper, and plastics. Persistence varies with degree of halogenation, an estimated half-life of 10 years. PCBs are toxic to fish at high doses, and associated with spawning failure at low doses. Human exposure occurs through food, and is associated with reproductive failure and immune suppression. Immediate effects of PCB exposure include pigmentation of nails and mucous membranes and swelling of the eyelids, along with fatigue, nausea, and vomiting. Effects are transgenerational, as the chemical can persist in a mother's body for up to 7 years, resulting in developmental delays and behavioral problems in her children. Food contamination has led to large scale PCB exposure.

10. Dichlorodiphenyltrichloroethane (DDT) is probably the most infamous POP. It was widely used as insecticide during WWII to protect against malaria and typhus. After the war, DDT was used as an agricultural insecticide. In 1962, the American biologist Rachel Carson published Silent Spring, describing the impact of DDT spraying on the US environment and human health. DDT's persistence in the soil for up to 10–15 years after application has resulted in wide-

spread and persistent DDT residues throughout the world including the arctic, even though it has been banned or severely restricted in most of the world. DDT is toxic to many organisms including birds where it is detrimental to reproduction due to eggshell thinning. DDT can be detected in foods from all over the world and food-borne DDT remains the greatest source of human exposure. Short-term acute effects of DDT on humans are limited, however long-term exposure has been associated with chronic health effects including increased risk of cancer and diabetes, reduced reproductive success, and neurological disease.

11. Dioxins are unintentional by-products of high-temperature processes, such as incomplete combustion and pesticide production. Dioxins are typically emitted from the burning of hospital waste, municipal waste, and hazardous waste, along with automobile emissions, peat, coal, and wood. Dioxins have been associated with several adverse effects in humans, including immune and enzyme disorders, chloracne, and are classified as a possible human carcinogen. In laboratory studies of dioxin effects an increase in birth defects and stillbirths, and lethal exposure have been associated with the substances. Food, particularly from animals, is the principal source of human exposure to dioxins.

12. Polychlorinated dibenzofurans are by-products of high-temperature processes, such as incomplete combustion after waste incineration or in automobiles, pesticide production, and polychlorinated biphenyl production. Structurally similar to dioxins, the two compounds share toxic effects. Furans persist in the environment and classified as possible human carcinogens. Human exposure to furans primarily results from food, particularly animal products.

New POPs on the Stockholm Convention List

Since 2001, this list has been expanded to include some polycyclic aromatic hydrocarbons (PAHs), brominated flame retardants, and other compounds. Additions to the initial 2001 Stockholm Convention list are as following POPs:

- Chlordecone, a synthetic chlorinated organic compound,is primarily used as an agricultural pesticide, related to DDT and Mirex. Chlordecone is toxic to aquatic organisms, and classified as a possible human carcinogen. Many countries have banned chlordecone sale and use, or intend to phase out stockpiles and wastes.

- α-Hexachlorocyclohexane (α-HCH) and β-Hexachlorocyclohexane (β-HCH) are insecticides as well as by-products in the production of lindane. Large stockpiles of HCH isomers exist in the environment. α-HCH and β-HCH are highly persistent in the water of colder regions. α-HCH and β-HCH has been linked Parkinson's and Alzheimer's disease.

- Hexabromodiphenyl ether (hexaBDE) and heptabromodiphenyl ether (heptaBDE) are main components of commercial octabromodiphenyl ether (octaBDE).

Commercial octaBDE is highly persistent in the environment, whose only degradation pathway is through debromination and the production of bromodiphenyl ethers, which can increase toxicity.

- Lindane (γ-hexachlorocyclohexane), a pesticide used as a broad spectrum insecticide for seed, soil, leaf, tree and wood treatment, and against ectoparasites in animals and humans (head lice and scabies). Lindane rapidly bioconcentrates. It is immunotoxic, neurotoxic, carcinogenic, linked to liver and kidney damage as well as adverse reproductive and developmental effects in laboratory animals and aquatic organisms. Production of lindane unintentionally produces two other POPs α-HCH and β-HCH.

- Pentachlorobenzene (PeCB), is a pesticide and unintentional byproduct. PeCB has also been used in PCB products, dyestuff carriers, as a fungicide, a flame retardant, and a chemical intermediate. PeCB is moderately toxic to humane, while highly toxic to aquatic organisms.

- Tetrabromodiphenyl ether (tetraBDE) and pentabromodiphenyl ether (pentaBDE) are industrial chemicals and the main components of commercial pentabromodiphenyl ether (pentaBDE). PentaBDE has been detected in humans in all regions of the world.

- Perfluorooctanesulfonic acid (PFOS) and its salts are used in the production of fluoropolymers. PFOS and related compounds are extremely persistent, bioaccumulating and biomagnifying. The negative effects of trace levels of PFOS have not been established.

- Endosulfans are insecticides to control pests on crops such coffee, cotton, rice and sorghum and soybeans, tsetse flies, ectoparasites of cattle. They are used as a wood preservative. Global use and manufacturing of endosulfan has been banned under the Stockholm convention in 2011, although many countries had previously banned or introduced phase-outs of the chemical when the ban was announced. Toxic to humans and aquatic and terrestrial organisms, linked to congenital physical disorders, mental retardation, and death. Endosulfans' negative health effects are primarily liked to its endocrine disrupting capacity acting as an antiandrogen.

- Hexabromocyclododecane (HBCD) is a brominated flame retardant primarily used in thermal insulation in the building industry. HBCD is persistent, toxic and ecotoxic, with bioaccumulative and long-range transport properties.

Additive and Synergistic Effects

Evaluation of the effects of POPs on health is very challenging in the laboratory setting. For example, for organisms exposed to a mixture of POPs, the effects are assumed to be additive. Mixtures of POPs can in principle produce synergistic effects. With synergistic effects, the toxicity of each compound is enhanced (or depressed) by the presence of other compounds in the mixture. When put together, the effects can far exceed the approximated additive effects of the POP compound mixture.

Health Effects

POP exposure may cause developmental defects, chronic illnesses, and death. Some are carcinogens per IARC, possibly including breast cancer. Many POPs are capable of endocrine disruption within the reproductive system, the central nervous system, or the immune system. People and animals are exposed to POPs mostly through their diet, occupationally, or while growing in the womb. For humans not exposed to POPs through accidental or occupational means, over 90% of exposure comes from animal product foods due to bioaccumulation in fat tissues and bioaccumulate through the food chain. In general, POP serum levels increase with age and tend to be higher in females than males.

Studies have investigated the correlation between low level exposure of POPs and various diseases. In order to assess disease risk due to POPs in a particular location, government agencies may produce a human health risk assessment which takes into account the pollutants' bioavailability and their dose-response relationships.

Endocrine Disruption

The majority of POPs are known to disrupt normal functioning of the endocrine system, for example all of the dirty dozen are endocrine disruptors. Low level exposure to POPs during critical developmental periods of fetus, newborn and child can have a lasting effect throughout its lifespan. A 2002 study synthesizes data on endocrine disruption and health complications from exposure to POPs during critical developmental stages in an organism's lifespan. The study aimed to answer the question whether or not chronic, low level exposure to POPs can have a health impact on the endocrine system and development of organisms from different species. The study found that exposure of POPs during a critical developmental time frame can produce a permanent changes in the organisms path of development. Exposure of POPs during non-critical developmental time frames may not lead to detectable diseases and health complications later in their life. In wildlife, the critical development time frames are in utero, in ovo, and during reproductive periods. In humans, the critical development timeframe is during fetal development.

Reproductive System

The same study in 2002 with evidence of a link from POPs to endocrine disruption also linked low dose exposure of POPs to reproductive health effects. The study stated that POP exposure can lead to negative health effects especially in the male reproductive system, such as decreased sperm quality and quantity, altered sex ratio and early puberty onset. For females exposed to POPs, altered reproductive tissues and pregnancy outcomes as well as endometriosis have been reported.

Exposure During Pregnancy

POP exposure during pregnancy is of particular concern to the developing fetus.

Transport Across the Placenta

A study about the transfer of POPs (14 organochlorine pesticides, 7 polychlorinated biphenyls and 14 polybrominated diphenyl ethers (PBDEs)) from Spanish mothers to their unborn fetus found that POP concentrations in serum from the mother were higher than from the umbilical cord and 50 placentas. Because transfer of the POPs from mother to fetus did not correspond with passive lipid-associated diffusion, authors suggested that POPs are actively transported across the placenta.

Gestational Weight Gain and Newborn Head Circumference

A Greek study from 2014 investigated the link between maternal weight gain during pregnancy, their PCB-exposure level and PCB level in their newborn infants, their birth weight, gestational age, and head circumference. The birth weight and head circumference of the infants was the lower, the higher POP levels during prenatal development had been, but only if mothers had either excessive or inadequate weight gain during pregnancy. No correlation between POP exposure and gestational age was found. A 2013 case-control study conducted 2009 in Indian mothers and their offspring showed prenatal exposure of two types of organochlorine pesticides (HCH, DDT and DDE) impaired the growth of the fetus, reduced the birth weight, length, head circumference and chest circumference.

Cardiovascular Disease and Cancer

POPs are lipophilic environmental toxins. They are often found in lipoproteins of organisms. A study published in 2014 found an association between the concentration of POPs in lipoproteins and the occurrence of cardiovascular disease and various cancers in human beings. The higher the concentration of POPs found in lipoproteins, the higher the occurrence of cardiovascular disease and cancer. Highly chlorinated polychlorinated biphenyls are specifically found in high concentrations in lipoproteins. Cardiovascular disease is shown to be more associated with higher concentrations of POPs in high density lipoproteins and cancer is shown to be more associated with higher concentrations of POPs in low density lipoproteins and very low density lipoproteins.

Obesity

There have been many recent studies assessing the connection between serum POP levels in individuals and instances of obesity. A study released in 2011 found correlations between different POPs and obesity occurrence in individuals tested. The statistically significant findings from the study show that there is actually a negative correlation between various PCB congener serum levels and obesity in individuals tested. The study also showed a positive correlation between beta-hexachlorocyclohexane and various dioxin serum levels and obesity in individuals tested. Obesity was determined using the Body Mass Index (BMI). One proposed explanation in the study is that PCBs are very lipophilic, therefore they are easily stored and captured in the fat deposits in human beings. Obese individuals have higher amounts of fat deposits in their body, and thus more PCBs could be captured in the fat deposits leading to less PCBs circulating in blood serum. The study provides evidence demonstrating that the correlation between POP serum levels and obesity occurrence is more complicated than previously expected. The same study also noted a strong positive correlation between serum POP levels and age in all individuals in the experiment.

Diabetes

A study published in 2006 revealed a positive correlation between POP serum levels and type II diabetes in individuals, after other variables, such as age, sex, race, and socioeconomic status were adjusted for. The correlation proved stronger in younger, Mexican American, and obese individuals. Individuals exposed to low doses of POPs throughout their lifetime had a higher chance for developing diabetes than individuals exposed to high concentrations of POPs for a short amount of time.

POPs in Urban Areas and Indoor Environments

Traditionally it was thought that human exposure to POPs occurred primarily through food, however indoor pollution patterns that characterize certain POPs have challenged this notion. Recent studies of indoor dust and air have implicated indoor environments as a major sources for human exposure via inhalation and ingestion. Furthermore, significant indoor POP pollution must be a major route of human POP exposure, considering the modern trend in spending larger proportions of life indoor. Several studies have shown that indoor (air and dust) POP levels to exceed outdoor (air and soil) POP concentrations.

Control and Removal of POPs in the Environment

Current studies aimed at minimizing POPs in the environment are investigating their behavior in photo catalytic oxidation reactions. POPs that are found in humans and in aquatic environments the most are the main subjects of these experiments. Aromatic and aliphatic degradation products have been identified in these reactions. Photochemical degradation is negligible compared to photocatalytic degradation. However, proper removal techniques of POPs from the environment are still unclear, due to fear that more toxic byproducts may result from uninvestigated degradation techniques. Current efforts are more focused on banning the use and production of POPs worldwide rather than removal of POPs.

2

Toxin Degradation Processes

The processes concerned within this text are chemical decomposition, biodegradation, photodissociation and photocatalysis. Chemical decomposition is the breakdown of chemical compounds into simpler compounds whereas the fragmentation of materials by bacteria or fungi is known as biodegradation. The major processes of toxic degradation are dealt with great details in the text. The aspects elucidated in this chapter are of vital importance, and provide a better understanding of persistent organic pollutants.

Chemical Decomposition

Chemical decomposition, analysis or breakdown is the separation of a chemical compound into elements or simpler compounds. It is sometimes defined as the exact opposite of a chemical synthesis. Chemical decomposition is often an undesired chemical reaction. The stability that a chemical compound ordinarily has is eventually limited when exposed to extreme environmental conditions like heat, radiation, humidity or the acidity of a solvent. The details of decomposition processes are generally not well defined, as a molecule may break up into a host of smaller fragments. Chemical decomposition is exploited in several analytical techniques, notably mass spectrometry, traditional gravimetric analysis, and thermogravimetric analysis.

A broader definition of the term decomposition also includes the breakdown of one phase into two or more phases.

There are three broad types of decomposition reactions: thermal, electrolytic and catalytic.

Reaction Formula

The generalized reaction for chemical decomposition is:

AB → A + B with a specific example being the electrolysis of water to gaseous hydrogen and oxygen:

$2 H_2O(l) \rightarrow 2 H_2 + O_2$

Additional Examples

An example of spontaneous decomposition is that of hydrogen peroxide, which will slowly decompose into water and oxygen:

$$2\ H_2O_2 \rightarrow 2\ H_2O + O_2$$

Carbonates will decompose when heated, a notable exception being that of carbonic acid, H_2CO_3. Carbonic acid, the "fizz" in sodas, pop cans and other carbonated beverages, will decompose over time (spontaneously) into carbon dioxide and water

$$H_2CO_3 \rightarrow H_2O + CO_2$$

Other carbonates will decompose when heated producing the corresponding metal oxide and carbon dioxide. In the following equation M represents a metal:

$$MCO_3 \rightarrow MO + CO_2$$

A specific example of this involving calcium carbonate:

$$CaCO_3 \rightarrow CaO + CO_2$$

Metal chlorates also decompose when heated. A metal chloride and oxygen gas are the products.

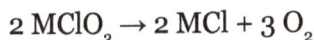

$$2\ MClO_3 \rightarrow 2\ MCl + 3\ O_2$$

A common decomposition of a chlorate to evolve oxygen utilizes potassium chlorate as follows:

$$2\ KClO_3 \rightarrow 2\ KCl + 3\ O_2$$

Biodegradation

Yellow slime mold growing on a bin of wet paper

IUPAC Definition

Biodegradation is the disintegration of materials by bacteria, fungi, or other biological means. Although often conflated, biodegradable is distinct in meaning from compostable. While biodegradable simply means to be consumed by microorganisms, "compostable" makes the specific demand that the object break down under composting conditions. The term is often used in relation to ecology, waste management, biomedicine, and the natural environment (bioremediation) and is now commonly associated with environmentally friendly products that are capable of decomposing back into natural elements. Organic material can be degraded aerobically with oxygen, or anaerobically, without oxygen. Biosurfactant, an extracellular surfactant secreted by microorganisms, enhances the biodegradation process.

Biodegradable matter is generally organic material that serves as a nutrient for microorganisms. Microorganisms are so numerous and diverse that, a huge range of compounds are biodegraded, including hydrocarbons (e.g. oil), polychlorinated biphenyls (PCBs), polyaromatic hydrocarbons (PAHs), pharmaceutical substances. Decomposition of biodegradable substances may include both biological and abiotic steps.

Factors Affecting Rate

In practice, almost all chemical compounds and materials are subject to biodegradation, the key is the relative rates of such processes - minutes, days, years, centuries... A number of factors determine the degradation rate of organic compounds. Salient factors include light, water and oxygen. Temperature is also important because chemical reactions proceed more quickly at higher temperatures. The degradation rate of many organic compounds is limited by their bioavailability. Compounds must be released into solution before organisms can degrade them.

Biodegradability can be measured in a number of ways. Respirometry tests can be used for aerobic microbes. First one places a solid waste sample in a container with microorganisms and soil, and then aerate the mixture. Over the course of several days, microorganisms digest the sample bit by bit and produce carbon dioxide – the resulting amount of CO_2 serves as an indicator of degradation. Biodegradability can also be measured by anaerobic microbes and the amount of methane or alloy that they are able to produce. In formal scientific literature, the process is termed bio-remediation.

| Approximated time for compounds to biodegrade in a marine environment ||
Product	Time to Biodegrade
Paper towel	2–4 weeks
Newspaper	6 weeks
Apple core	2 months
Cardboard box	2 months
Wax coated milk carton	3 months
Cotton gloves	1–5 months
Wool gloves	1 year
Plywood	1–3 years

Painted wooden sticks	13 years
Plastic bags	10–20 years
Tin cans	50 years
Disposable diapers	50–100 years
Plastic bottle	100 years
Aluminium cans	200 years
Glass bottles	Undetermined

Detergents

In advanced societies, laundry detergents are based on *linear* alkylbenzenesulfonates. Branched alkybenzenesulfonates (below right), used in former times, were abandoned because they biodegrade too slowly.

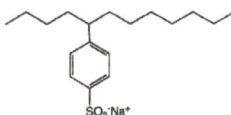

4-(5-Dodecyl) benzenesulfonate, a linear dodecylbenzenesulfonate

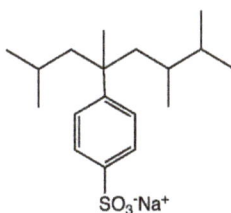

A branched dodecylbenzenesulfonate, which has been phased out in developed countries.

Plastics

Plastics biodegrade at highly variable rates. PVC-based plumbing is specifically selected for handing sewage because PVC biodegrades very slowly. Some packaging materials on the other hand are being developed that would degrade readily upon exposure to the environment. Illustrative synthetic polymers that are biodegrade quickly include polycaprolactone, others are polyesters and aromatic-aliphatic esters, due to their ester bonds being susceptible to attack by water. A prominent example is poly-3-hydroxybutyrate, the renewably derived polylactic acid, and the synthetic polycaprolactone. Others are the cellulose-based cellulose acetate and celluloid (cellulose nitrate).

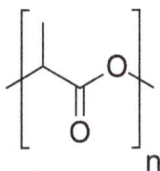

Polylactic acid is an example of a plastic that biodegrades quickly.

Under low oxygen conditions biodegradable plastics break down slower and with the production of methane, like other organic materials do. The breakdown process is accelerated in a dedicated compost heap. Starch-based plastics will degrade within two to four months in a home compost bin, while polylactic acid is largely undecomposed, requiring higher temperatures. Polycaprolactone and polycaprolactone-starch composites decompose slower, but the starch content accelerates decomposition by leaving behind a porous, high surface area polycaprolactone. Nevertheless, it takes many months. In 2016, a bacterium named Ideonella sakaiensis was found to biodegrade PET.

Many plastic producers have gone so far even to say that their plastics are compostable, typically listing corn starch as an ingredient. However, these claims are questionable because the plastics industry operates under its own definition of compostable:

> "that which is capable of undergoing biological decomposition in a compost site such that the material is not visually distinguishable and breaks down into carbon dioxide, water, inorganic compounds and biomass at a rate consistent with known compostable materials."

The term "composting" is often used informally to describe the biodegradation of packaging materials. Legal definitions exist for compostability, the process that leads to compost. Four criteria are offered by the European Union:

- Biodegradability, the conversion of >90% material material into CO2 and water by the action of micro-organisms within 6 months.

- Disintegrability, the fragmentation of 90% of the original mass to particles that then pass through a 2 mm sieve.

- Absence of toxic substances and other substances that impede composting.

Biodegradable Technology

In 1973 it was proven for the first time that polyester degrades when disposed in bioactive material such as soil. Polyesters are water resistant and can be melted and shaped into sheets, bottles, and other products, making certain plastics now available as a biodegradable product. Following, Polyhydroxylalkanoates (PHAs) were produced directly from renewable resources by microbes. They are approximately 95% cellular bacteria and can be manipulated by genetic strategies. The composition and biodegradability of PHAs can be regulated by blending it with other natural polymers. In the 1980s the company ICI Zenecca commercialized PHAs under the name Biopol. It was used for the production of shampoo bottles and other cosmetic products. Consumer response was unusual. Consumers were willing to pay more for this product because it was natural and biodegradable, which had not occurred before.

Now biodegradable technology is a highly developed market with applications in product packaging, production and medicine. Biodegradable technology is concerned with the manufacturing science of biodegradable materials. It imposes science based mechanisms of plant genetics into the processes of today. Scientists and manufacturing corporations can help impact climate change by developing a use of plant genetics that would mimic some technologies. By looking to plants, such as biodegradable material harvested through photosynthesis, waste and toxins can be minimized.

Oxo-biodegradable technology, which has further developed biodegradable plastics, has also emerged. Oxo-biodegradation is defined by CEN (the European Standards Organisation) as "degradation resulting from oxidative and cell-mediated phenomena, either simultaneously or successively." Whilst sometimes described as "oxo-fragmentable," and "oxo-degradable" this describes only the first or oxidative phase. These descriptions should not be used for material which degrades by the process of oxo-biodegradation defined by CEN, and the correct description is "oxo-biodegradable."

By combining plastic products with very large polymer molecules, which contain only carbon and hydrogen, with oxygen in the air, the product is rendered capable of decomposing in anywhere from a week to one to two years. This reaction occurs even without prodegradant additives but at a very slow rate. That is why conventional plastics, when discarded, persist for a long time in the environment. Oxo-biodegradable formulations catalyze and accelerate the biodegradation process but it takes considerable skill and experience to balance the ingredients within the formulations so as to provide the product with a useful life for a set period, followed by degradation and biodegradation.

Biodegradable technology is especially utilized by the bio-medical community. Biodegradable polymers are classified into three groups: medical, ecological, and dual application, while in terms of origin they are divided into two groups: natural and synthetic. The Clean Technology Group is exploiting the use of supercritical carbon dioxide, which under high pressure at room temperature is a solvent that can use biodegradable plastics to make polymer drug coatings. The polymer (meaning a material composed of molecules with repeating structural units that form a long chain) is used to encapsulate a drug prior to injection in the body and is based on lactic acid, a compound normally produced in the body, and is thus able to be excreted naturally. The coating is designed for controlled release over a period of time, reducing the number of injections required and maximizing the therapeutic benefit. Professor Steve Howdle states that biodegradable polymers are particularly attractive for use in drug delivery, as once introduced into the body they require no retrieval or further manipulation and are degraded into soluble, non-toxic by-products. Different polymers degrade at different rates within the body and therefore polymer selection can be tailored to achieve desired release rates.

Other biomedical applications include the use of biodegradable, elastic shape-memory polymers. Biodegradable implant materials can now be used for minimally invasive surgical procedures through degradable thermoplastic polymers. These polymers are now able to change their shape with increase of temperature, causing shape memory capabilities as well as easily degradable sutures. As a result, implants can now fit through small incisions, doctors can easily perform complex deformations, and sutures and other material aides can naturally biodegrade after a completed surgery.

Etymology of "Biodegradable"

The first known use of the word in biological text was in 1961 when employed to describe the breakdown of material into the base components of carbon, hydrogen, and oxygen by microorganisms. Now biodegradable is commonly associated with environmentally friendly products that are part of the earth's innate cycle and capable of decomposing back into natural elements.

Photodissociation

Photodissociation, photolysis, or photodecomposition is a chemical reaction in which a chemical compound is broken down by photons. It is defined as the interaction of one or more photons with one target molecule. Photodissociation is not limited to visible light. Any photon with sufficient energy can affect the chemical bonds of a chemical compound. Since a photon's energy is inversely proportional to its wavelength, electromagnetic waves with the energy of visible light or higher, such as ultraviolet light, x-rays and gamma rays are usually involved in such reactions.

Photolysis in Photosynthesis

Photolysis is part of the light-dependent reactions of photosynthesis. The general reaction of photosynthetic photolysis can be given as

$$H_2A + 2 \text{ photons (light)} \rightarrow 2 \text{ e}^- + 2 \text{ H}^+ + A$$

The chemical nature of "A" depends on the type of organism. In purple sulfur bacteria, hydrogen sulfide (H_2S) is oxidized to sulfur (S). In oxygenic photosynthesis, water (H_2O) serves as a substrate for photolysis resulting in the generation of diatomic oxygen (O_2). This is the process which returns oxygen to Earth's atmosphere. Photolysis of water occurs in the thylakoids of cyanobacteria and the chloroplasts of green algae and plants.

Energy Transfer Models

The conventional, semi-classical, model describes the photosynthetic energy transfer process as one in which excitation energy hops from light-capturing pigment molecules to reaction center molecules step-by-step down the molecular energy ladder.

The effectiveness of photons of different wavelengths depends on the absorption spectra of the photosynthetic pigments in the organism. Chlorophylls absorb light in the violet-blue and red parts of the spectrum, while accessory pigments capture other wavelengths as well. The phycobilins of red algae absorb blue-green light which penetrates deeper into water than red light, enabling them to photosynthesize in deep waters. Each absorbed photon causes the formation of an exciton (an electron excited to a higher energy state) in the pigment molecule. The energy of the exciton is transferred to a chlorophyll molecule (P680, where P stands for pigment and 680 for its absorption maximum at 680 nm) in the reaction center of photosystem II via resonance energy transfer. P680 can also directly absorb a photon at a suitable wavelength.

Photolysis during photosynthesis occurs in a series of light-driven oxidation events. The energized electron (exciton) of P680 is captured by a primary electron acceptor of the photosynthetic electron transfer chain and thus exits photosystem II. In order to repeat the reaction, the electron in the reaction center needs to be replenished. This occurs by oxidation of water in the case of oxygenic photosynthesis. The electron-deficient reaction center of photosystem II (P680*) is the strongest biological oxidizing agent yet discovered, which allows it to break apart molecules as stable as water.

The water-splitting reaction is catalyzed by the oxygen evolving complex of photosystem II. This protein-bound inorganic complex contains four manganese ions, plus calcium and chloride ions as cofactors. Two water molecules are complexed by the manganese cluster, which then undergoes a series of four electron removals (oxidations) to replenish the reaction center of photosystem II. At the end of this cycle, free oxygen (O_2) is generated and the hydrogen of the water molecules has been converted to four protons released into the thylakoid lumen.

These protons, as well as additional protons pumped across the thylakoid membrane coupled with the electron transfer chain, form a proton gradient across the membrane that drives photophosphorylation and thus the generation of chemical energy in the form of adenosine triphosphate (ATP). The electrons reach the P700 reaction center of photosystem I where they are energized again by light. They are passed down another electron transfer chain and finally combine with the coenzyme $NADP^+$ and protons outside the thylakoids to NADPH. Thus, the net oxidation reaction of water photolysis can be written as:

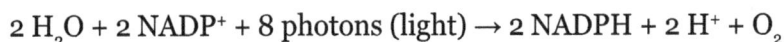

$$2\ H_2O + 2\ NADP^+ + 8\ \text{photons (light)} \rightarrow 2\ NADPH + 2\ H^+ + O_2$$

The free energy change (ΔG) for this reaction is 102 kilocalories per mole. Since the energy of light at 700 nm is about 40 kilocalories per mole of photons, approximately 320 kilocalories of light energy are available for the reaction. Therefore, approximately one-third of the available light energy is captured as NADPH during photolysis and electron transfer. An equal amount of ATP is generated by the resulting proton gradient. Oxygen as a byproduct is of no further use to the reaction and thus released into the atmosphere.

Quantum Models

In 2007 a quantum model was proposed by Graham Fleming and his co-workers which includes the possibility that photosynthetic energy transfer might involve quantum oscillations, explaining its unusually high efficiency.

According to Fleming there is direct evidence that remarkably long-lived wavelike electronic quantum coherence plays an important part in energy transfer processes during photosynthesis, which can explain the extreme efficiency of the energy transfer because it enables the system to sample all the potential energy pathways, with low loss, and choose the most efficient one. This claim has, however, since been proven wrong in several publications.

This approach has been further investigated by Gregory Scholes and his team at the University of Toronto, which in early 2010 published research results that indicate that some marine algae make use of quantum-coherent electronic energy transfer (EET) to enhance the efficiency of their energy harnessing.

Photolysis in The Atmosphere

Photolysis occurs in the atmosphere as part of a series of reactions by which primary pollutants such as hydrocarbons and nitrogen oxides react to form secondary pollut-ants such as peroxyacyl nitrates.

The two most important photodissociaton reactions in the troposphere are firstly:

$$O_3 + h\nu \rightarrow O_2 + O(^1D) \quad \lambda < 320 \text{ nm}$$

which generates an excited oxygen atom which can react with water to give the hydroxyl radical:

$$O(^1D) + H_2O \rightarrow 2 \; {}^{\cdot}OH$$

The hydroxyl radical is central to atmospheric chemistry as it initiates the oxidation of hydrocarbons in the atmosphere and so acts as a detergent.

Secondly the reaction:

$$NO_2 + h\nu \rightarrow NO + O$$

is a key reaction in the formation of tropospheric ozone.

The formation of the ozone layer is also caused by photodissociation. Ozone in the Earth's stratosphere is created by ultraviolet light striking oxygen molecules containing two oxygen atoms (O_2), splitting them into individual oxygen atoms (atomic oxygen). The atomic oxygen then combines with unbroken O_2 to create ozone, O_3. In addition, photolysis is the process by which CFCs are broken down in the upper atmosphere to form ozone-destroying chlorine free radicals.

Astrophysics

In astrophysics, photodissociation is one of the major processes through which molecules are broken down (but new molecules are being formed). Because of the vacuum of the interstellar medium, molecules and free radicals can exist for a long time. Photodissociation is the main path by which molecules are broken down. Photodissociation rates are important in the study of the composition of interstellar clouds in which stars are formed.

Examples of photodissociation in the interstellar medium are is the energy of a single photon of frequency

Atmospheric Gamma-ray Bursts

Currently orbiting satellites detect an average of about one gamma-ray burst per day. Because gamma-ray bursts are visible to distances encompassing most of the observable universe, a volume encompassing many billions of galaxies, this suggests that gamma-ray bursts must be exceedingly rare events per galaxy.

Measuring the exact rate of gamma-ray bursts is difficult, but for a galaxy of approximately the same size as the Milky Way, the expected rate (for long GRBs) is about one burst every 100,000 to 1,000,000 years. Only a few percent of these would be beamed towards Earth. Estimates of rates of short GRBs are even more uncertain because of the unknown beaming fraction, but are probably comparable.

A gamma-ray burst in the Milky Way, if close enough to Earth and beamed towards it, could have significant effects on the biosphere. The absorption of radiation in the atmosphere would cause photodissociation of nitrogen, generating nitric oxide that would act as a catalyst to destroy ozone.

The atmospheric photodissociation

- $N_2 \rightarrow 2N$
- $O_2 \rightarrow 2O$
- $CO_2 \rightarrow C + 2O$
- $H_2O \rightarrow 2H + O$
- $2NH_3 \rightarrow 3H_2 + N_2$

would yield

- NO_2 (consumes up to 400 ozone molecules)
- CH_2 (nominal)

- CH_4 (nominal)

- CO_2

According to a 2004 study, a GRB at a distance of about a kiloparsec could destroy up to half of Earth's ozone layer; the direct UV irradiation from the burst combined with additional solar UV radiation passing through the diminished ozone layer could then have potentially significant impacts on the food chain and potentially trigger a mass extinction. The authors estimate that one such burst is expected per billion years, and hypothesize that the Ordovician-Silurian extinction event could have been the result of such a burst.

There are strong indications that long gamma-ray bursts preferentially or exclusively occur in regions of low metallicity. Because the Milky Way has been metal-rich since before the Earth formed, this effect may diminish or even eliminate the possibility that a long gamma-ray burst has occurred within the Milky Way within the past billion years. No such metallicity biases are known for short gamma-ray bursts. Thus, depending on their local rate and beaming properties, the possibility for a nearby event to have had a large impact on Earth at some point in geological time may still be significant.

Multiple Photon Dissociation

Single photons in the infrared spectral range usually are not energetic enough for direct photodissociation of molecules. However, after absorption of multiple infrared photons a molecule may gain internal energy to overcome its barrier for dissociation. Multiple photon dissociation (MPD, IRMPD with infrared radiation) can be achieved by applying high power lasers, e.g. a carbon dioxide laser, or a free electron laser, or by long interaction times of the molecule with the radiation field without the possibility for rapid cooling, e.g. by collisions. The latter method allows even for MPD induced by black body radiation, a technique called blackbody infrared radiative dissociation (BIRD).

Photocatalysis

In chemistry, photocatalysis is the acceleration of a photoreaction in the presence of a catalyst. In catalysed photolysis, light is absorbed by an adsorbed substrate. In photogenerated catalysis, the photocatalytic activity (PCA) depends on the ability of the catalyst to create electron–hole pairs, which generate free radicals (e.g. hydroxyl radicals: •OH) able to undergo secondary reactions. Its practical application was made possible by the discovery of water electrolysis by means of titanium dioxide. The commercially used process is called the advanced oxidation process (AOP). There are several ways the AOP can be carried out; these may (but do not necessarily) involve TiO_2 or even the use of UV light. Generally the defining factor is the production and use of the hydroxyl radical.

In the experiment above, photons from a light source (out of frame on the right hand side) are absorbed by the surface of the titanium dioxide disc, exciting electrons within the material. These then react with the water molecules, splitting it into its constituents of hydrogen and oxygen. In this experiment, chemicals dissolved in the water prevent the formation of oxygen, which would otherwise recombine with the hydrogen.

Types of Photocatalysis

Homogeneous Photocatalysis

In homogeneous photocatalysis, the reactants and the photocatalysts exist in the same phase. The most commonly used homogeneous photocatalysts include ozone and photo-Fenton systems (Fe^+ and Fe^+/H_2O_2). The reactive species is the •OH which is used for different purposes. The mechanism of hydroxyl radical production by ozone can follow two paths.

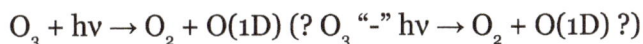

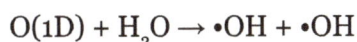

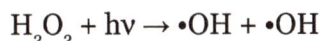

$$O_3 + h\nu \rightarrow O_2 + O(1D) \ (? \ O_3 \ \text{"-"} \ h\nu \rightarrow O_2 + O(1D) \ ?)$$

$$O(1D) + H_2O \rightarrow \bullet OH + \bullet OH$$

$$O(1D) + H_2O \rightarrow H_2O_2$$

$$H_2O_2 + h\nu \rightarrow \bullet OH + \bullet OH$$

Similarly, the Fenton system produces hydroxyl radicals by the following mechanism

$$Fe^{2+} + H_2O_2 \rightarrow HO\bullet + Fe^{3+} + OH^-$$

$$Fe^{3+} + H_2O_2 \rightarrow Fe^{2+} + HO \bullet 2 + H^+$$

$$Fe^{2+} + HO \bullet \rightarrow Fe^{3+} + OH^-$$

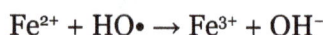

In photo-Fenton type processes, additional sources of OH radicals should be considered: through photolysis of H_2O_2, and through reduction of Fe^{3+} ions under UV light:

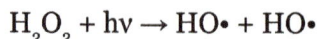

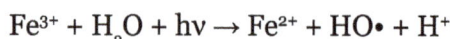

$$H_2O_2 + h\nu \rightarrow HO \bullet + HO \bullet$$

$$Fe^{3+} + H_2O + h\nu \rightarrow Fe^{2+} + HO \bullet + H^+$$

The efficiency of Fenton type processes is influenced by several operating parameters like concentration of hydrogen peroxide, pH and intensity of UV. The main advantage of this process is the ability of using sunlight with light sensitivity up to 450 nm, thus avoiding the high costs of UV lamps and electrical energy. These reactions have been proven more efficient than the other photocatalysis but the disadvantages of the process are the low pH values which are required, since iron precipitates at higher pH values and the fact that iron has to be removed after treatment.

Heterogeneous Photocatalysis

Heterogeneous catalysis has the catalyst in a different phase from the reactants. Heterogeneous photocatalysis is a discipline which includes a large variety of reactions: mild or total oxidations, dehydrogenation, hydrogen transfer, $^{18}O_2 - ^{16}O_2$ and deuterium-alkane isotopic exchange, metal deposition, water detoxification, gaseous pollutant removal, etc.

Most common heterogeneous photocatalysts are transition metal oxides and semiconductors, which have unique characteristics. Unlike the metals which have a continuum of electronic states, semiconductors possess a void energy region where no energy levels are available to promote recombination of an electron and hole produced by photoactivation in the solid. The void region, which extends from the top of the filled valence band to the bottom of the vacant conduction band, is called the band gap. When a photon with energy equal to or greater than the materials band gap is absorbed by the semiconductor, an electron is excited from the valence band to the conduction band, generating a positive hole in the valence band. The excited electron and hole can recombine and release the energy gained from the excitation of the electron as heat. Recombination is undesirable and leads to an inefficient photocatalyst. The ultimate goal of the process is to have a reaction between the excited electrons with an oxidant to produce a reduced product, and also a reaction between the generated holes with a reductant to produce an oxidized product. Due to the generation of positive holes and electrons, oxidation-reduction reactions take place at the surface of semiconductors. In the oxidative reaction, the positive holes react with the moisture present on the surface and produce a hydroxyl radical.

Oxidative reactions due to photocatalytic effect:

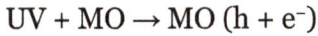

$$UV + MO \rightarrow MO (h + e^-)$$

Here MO stands for metal oxide ---

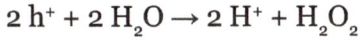

$$h^+ + H_2O \rightarrow H^+ + \bullet OH$$

$$2\,h^+ + 2\,H_2O \rightarrow 2\,H^+ + H_2O_2$$

$$H_2O_2 \rightarrow 2\,\bullet OH$$

The reductive reaction due to photocatalytic effect:

$$e^- + O_2 \rightarrow \bullet O_2^-$$

$$\bullet O_2^- + HO\bullet 2 + H^+ \rightarrow H_2O_2 + O_2$$

$$HOOH \rightarrow HO\bullet + \bullet OH$$

Ultimately, the hydroxyl radicals are generated in both the reactions. These hydroxyl radicals are very oxidative in nature and non selective with redox potential of ($E_o = +3.06\ V$)

Applications

- Conversion of water to hydrogen gas by photocatalytic water splitting. An efficient photocatalyst in the UV range is based on a sodium tantalite ($NaTaO_3$) doped with La and loaded with a cocatalyst nickel oxide. The surface of the sodium tantalite crystals is grooved with so called nanosteps that is a result of doping with lanthanum (3–15 nm range). The NiO parti-cles which facilitate hydrogen gas evolution are present on the edges, with the oxygen gas evolving from the grooves.

- Use of titanium dioxide in self-cleaning glass. Free radicals generated from TiO_2 oxidize organic matter.

- Disinfection of water by supported titanium dioxide photocatalysts, a form of solar water disinfection (SODIS).

- Use of titanium dioxide in self-sterilizing photocatalytic coatings (for application to food contact surfaces and in other environments where microbial pathogens spread by indirect contact).

- Oxidation of organic contaminants using magnetic particles that are coated with titanium dioxide nanoparticles and agitated using a magnetic field while being exposed to UV light.

- Conversion of carbon dioxide into gaseous hydrocarbons using titanium dioxide in the presence of water. As an efficient absorber in the UV range, titanium dioxide nanoparticles in the anatase and rutile phases are able to generate excitons by promoting electrons across the band gap. The electrons and holes react with the surrounding water vapor to produce hydroxyl radicals and protons. At present, proposed reaction mechanisms usually suggest the creation of a highly reactive carbon radical from carbon monoxide and carbon dioxide which then reacts with the photogenerated protons to ultimately form methane. Although the efficiencies of present titanium dioxide based photocatalysts are low, the incorporation of carbon based nanostructures such as carbon nanotubes and metallic nanoparticles have been shown to enhance the efficiency of these photocatalysts.

- Sterilization of surgical instruments and removal of unwanted fingerprints from sensitive electrical and optical components.

- A less-toxic alternative to tin and copper-based antifouling marine paints, ePaint, generates hydrogen peroxide by photocatalysis.

- Decomposition of crude oil with TiO_2 nanoparticles: by using titanium dioxide photocatalysts and UV-A radiation from the sun, the hydrocarbons found in crude oil can be turned into H_2O and CO_2. Higher amounts of oxygen and UV radiation increased the degradation of the model organics. These particles can be placed on floating substrates, making it easier to recover and catalyze the reaction. This is relevant since oil slicks float on top of the ocean and photons from the sun target the surface more than the inner depth of the ocean. By covering floating substrates like woodchips with epoxy adhesives, water logging can be prevented and TiO_2 particles can stick to the substrates. With more research, this method should be applicable to other organics.

- Decontamination of water with photocatalysis and adsorption: the removal and destruction of organic contaminants in groundwater can be addressed through the impregnation of adsorbents with photoactive catalysts. These adsorbents attract contaminating organic atoms/molecules like tetrachloroethylene to them. The photoactive catalysts impregnated inside speed up the degradation of the organics. Adsorbents are placed in packed beds for 18 hours, which would attract and degrade the organic compounds. The spent adsorbents would then be placed in regeneration fluid, essentially taking away all organics still attached by passing hot water counter-current to the flow of water during the adsorption process to speed up the reaction. The regeneration fluid then gets passed through the fixed beds of silica gel photocatalysts to remove and decompose the rest of the organics left. Through the use of fixed bed reactors, the regeneration of adsorbents can help increase the efficiency.

- Decomposition of polyaromatic hydrocarbons (PAHs). Triethylamine (TEA) was utilized to solvate and extract the polyaromatic hydrocarbons (PAHs) found in crude oil. By solvating these PAHs, TEA can attract the PAHs to itself. Once removed, TiO_2 slurries and UV light can photocatalytically degrade the PAHs. The figure shows the high success rate of this experiment. With high yielding of recoveries of 93–99% of these contaminants, this process has become an innovative idea that can be finalized for actual environmental usage. This procedure demonstrates the ability to develop photocatalysts that would be performed at ambient pressure, ambient temperature, and at a cheaper cost.

Quantification of Photocatalytic Activity

ISO 22197 -1:2007 specifies a test method for the determination of the nitric oxide removal performance of materials that contain a photocatalyst or have photocatalytic films on the surface.

Specific FTIR systems are used to characterise photocatalytic activity and/or passivity especially with respect to Volatile Organic Compounds VOCs and representative matrices of the binders applied.

Recent studies show that mass spectrometry can be a powerful tool to determine photocatalytic activity of certain materials by following the decomposition of gaseous pollutants such as nitrogen oxides or carbon dioxide

References

- Campbell, Neil A.; Reece, Jane B. (2005). Biology (7th ed.). San Francisco: Pearson - Benjamin Cummings. pp. 186–191. ISBN 0-8053-7171-0.

- Raven, Peter H.; Ray F. Evert; Susan E. Eichhorn (2005). Biology of Plants (7th ed.). New York: W.H. Freeman and Company Publishers. pp. 115–127. ISBN 0-7167-1007-2.

- Kyrikou, Ioanna; Briassoulis, Demetres (12 Apr 2007). "Biodegradation of Agricultural Plastic Films: A Critical Review". Journal of Polymers and the Environment. SpringerLink . 15 (2): 125–150. doi:10.1007/s10924-007-0053-8. Retrieved 30 May 2015.

3

Organobromine and Organochloride Compounds

The compounds that generate organic pollutants are organobromines, flame-retardants, organochlorides and dioxins and dioxin-like compounds. Organochlorides and organochlorine compounds contain at least one atom of chlorine. Flame retardants are the chemical substances that are added to either plastic or textiles. The aspects elucidated in this chapter are of vital importance, and provide a better understanding of organobromine and organochloride compounds.

Organobromine Compound

Organobromine compounds, also called *organobromides*, are organic compounds that contain carbon bonded to bromine. The most pervasive is the naturally produced bromomethane. One prominent application is the use of polybrominated diphenyl ethers as fire-retardants. A variety of minor organobromine compounds are found in nature, but none are biosynthesized or required by mammals. Organobromine compounds have fallen under increased scrutiny for their environmental impact.

General Properties

Most organobromine compounds, like most organohalide compounds, are relatively nonpolar. Bromine is more electronegative than carbon (2.8 vs 2.5). Consequently, the carbon in a carbon–bromine bond is electrophilic, i.e. alkyl bromides are alkylating agents.

Carbon–halogen bond strengths, or bond dissociation energies are of 115, 83.7, 72.1, and 57.6 kcal/mol for bonded to fluorine, chlorine, bromine, or iodine, respectively.

The reactivity of organobromine compounds resembles but is intermediate between the reactivity of organochlorine and organoiodine compounds. For many applications, organobromides represent a compromise of reactivity and cost. The principal reactions for organobromides include dehydrobromination, Grignard reactions, reductive coupling, and nucleophilic substitution.

Synthetic Methods

From Bromine

Alkenes reliably add bromine without catalysis to give the vicinal dibromides:

$$RCH=CH_2 + Br_2 \rightarrow RCHBrCH_2Br$$

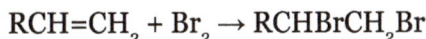

Aromatic compounds undergo bromination simultaneously with evolution of hydrogen bromide. Catalysts such as AlBr3 or FeBr3 are needed for the reaction to happen on aromatic rings. Chlorine-based catalysts (FeCl3, AlCl3) could be used, but yield would drop slightly as dihalogens(BrCl) could form. The reaction details following the usual patterns of electrophilic aromatic substitution:

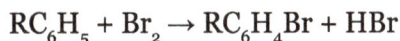

$$RC_6H_5 + Br_2 \rightarrow RC_6H_4Br + HBr$$

A prominent application of this reaction is the production of tetrabromobisphenol-A from bisphenol-A.

Free-radical substitution with bromine is commonly used to prepare organobromine compounds. Carbonyl-containing, benzylic, allylic substrates are especially prone to this reactions. For example, the commercially significant bromoacetic acid is generated directly from acetic acid and bromine in the presence of phosphorus tribromide catalyst:

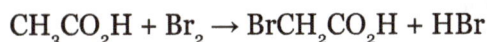

$$CH_3CO_2H + Br_2 \rightarrow BrCH_2CO_2H + HBr$$

Bromine also converts fluoroform to bromotrifluoromethane.

From Hydrogen Bromide

Hydrogen bromide adds across double bonds to give alkyl bromides, following the Markovnikov rule:

$$RCH=CH_2 + HBr \rightarrow RCHBrCH_3$$

Under free radical conditions, the direction of the addition can be reversed. Free-radical addition is used commercially for the synthesis of 1-bromoalkanes, precursors to tertiary amines and quaternary ammonium salts. 2-Phenethyl bromide ($C_6H_5CH_2CH_2Br$) is produced via this route from styrene.

Hydrogen bromide can also be used to convert alcohols to alkyl bromides. This reaction, that must be done under low temperature conditions, is employed in the industrial synthesis of allyl bromide:

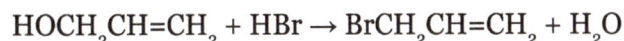

$$HOCH_2CH=CH_2 + HBr \rightarrow BrCH_2CH=CH_2 + H_2O$$

Methyl bromide, another fumigant, is generated from methanol and hydrogen bromide.

From Bromide Salts

Bromide ions, as provided by salts like sodium bromide, function as a nucleophiles in the formation of organobromine compounds by displacement.

An example of this salt mediated bromide displacement is the use of Copper(II) bromide on ketones:

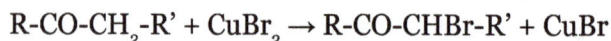

$$R\text{-}CO\text{-}CH_2\text{-}R' + CuBr_2 \rightarrow R\text{-}CO\text{-}CHBr\text{-}R' + CuBr$$

Industrially Significant Organobromine Compounds

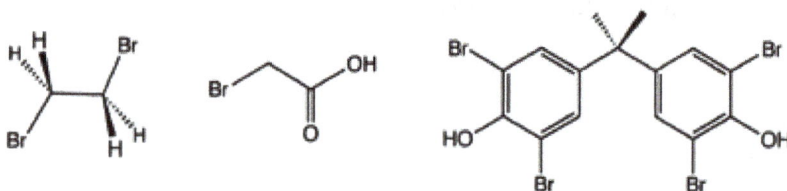

Structure of three industrially significant organobromine compounds. From left: ethylene bromide, bromoacetic acid, and tetrabromobisphenol-A.

Fire-retardants

Organobromine compounds are widely used as fire-retardants. The most prominent member is tetrabromobisphenol-A. It and tetrabromophthalic anhydride are precursors to polymers wherein the backbone features covalent carbon-bromine bonds. Other fire retardants, such as hexabromocyclododecane and the bromodiphenyl ethers, are additives and are not chemically attached to the material they protect. The use of organobromine fire-retardants is growing but is also controversial because they are persistent pollutants.

Fumigants and Biocides

Ethylene bromide, obtained by addition of bromine to ethylene, was once of commercial significance as a component of leaded gasoline. It was also a popular fumigant in agriculture, displacing 1,2-dibromo-3-chloropropane ("DBCB"). Both applications are declining owing to environmental and health considerations. Methyl bromide is also an effective fumigant, but its production and use are controlled by the Montreal Protocol. Growing in use are organobromine biocides used in water treatment. Representative agents include bromoform and dibromodimethylhydantoin ("DBDMH"). Some herbicides, such as bromoxynil, contain also bromine moieties. Like other halogenated pesticides, bromoxynil is subject to reductive dehalogenation under anaerobic conditions, and can be debrominated by organisms originally isolated for their ability to reductively dechlorinate phenolic compounds.

Dyes

Many dyes contain carbon-bromine bonds. The naturally occurring Tyrian purple (6,6'-dibromoindigo) was a valued dye before the development of the synthetic dye industry in the late 19th century. Several brominated anthroquinone derivatives are used commercially. Bromothymol blue is a popular indicator in analytical chemistry.

Pharmaceuticals

Commercially available organobromine pharmaceuticals include the vasodilator nicergoline, the sedative brotizolam, the anticancer agent pipobroman, and the antiseptic merbromin. Otherwise, organobromine compounds are rarely pharmaceutically useful, in contrast to the situation for organofluorine compounds. Several drugs are produced as the bromide (or equivalents, hydrobromide) salts, but in such cases bromide serves as an innocuous counterion of no biological significance.

Designer Drugs

Organobromine compounds such as 4-bromomethcathinone have appeared on the designer drug market alongside other halogenated amphetamines and cathinones in an attempt to circumvent existing drug laws.

Organobromine Compounds in Nature

Organobromine compounds are the most common organohalides in nature. Even though the concentration of bromide is only 0.3% of that for chloride in sea water, organobromine compounds are more prevalent in marine organisms than organochlorine derivatives. Their abundance reflects the easy oxidation of bromide to the equivalent of Br^+, a potent electrophile. The enzyme bromoperoxidase catalyzes this reaction. The oceans are estimated to release 1–2 million tons of bromoform and 56,000 tons of bromomethane annually. Red algae, such as the edible *Asparagopsis taxiformis*, eaten in Hawaii as "limu kohu", concentrate organobromine and organoiodine compounds in "vesicle cells"; 95% of the essential volatile oil of *Asparagopsis*, prepared by drying the seaweed in a vacuum and condensing using dry ice, is organohalogen compounds, of which bromoform comprises 80% by weight. Bromoform, produced by several algae, is a known toxin, though the small amounts present in edible algae do not appear to pose human harm. Some of these organobromine compounds are employed in a form of interspecies "chemical warfare." 5-Bromouracil and 3-Bromo-tyrosine have been identified in human white blood cells as products of myeloperoxidase-induced halogenation on invading pathogens.

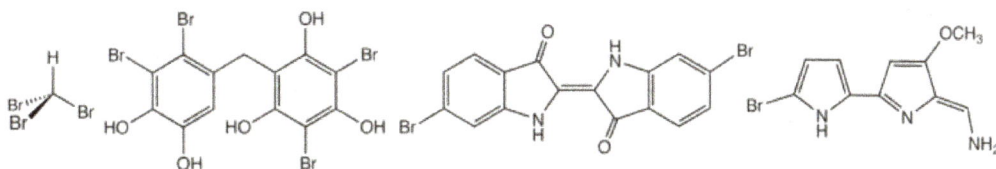

Structure of some naturally-occurring organobromine compounds. From left: bromoform, a brominated bisphenol, dibromoindigo (Tyrian purple), and the antifeedant tambjamine B.

In addition to conventional brominated natural products, a variety of organobromine compounds result from the biodegradation of fire-retardants. Metabolites include methoxylated and hydroxylated aryl bromides as well as brominated dioxin derivatives. Such compounds are considered persistent organic pollutants and have been found in mammals.

Safety

Alkyl bromine compounds are often alkylating agents and the brominated aromatic derivatives are implicated as hormone disruptors. Of the commonly produced compounds, ethylene dibromide is of greatest concern as it is both highly toxic and highly carcinogenic.

Flame Retardant

The term Flame retardants subsumes a diverse group of chemicals which are added to manufactured materials, such as plastics and textiles, and surface finishes and coatings. Flame retardants inhibit or delay the spread of fire by suppressing the chemical reactions in the flame or by the formation of a protective layer on the surface of a material. They may be mixed with the base material (additive flame retardants) or chemically bonded to it (reactive flame retardants). Mineral flame retardants are typically additive while organohalogen and organophosphorus compounds can be either reactive or additive.

Classes

Both Reactive and Additive Flame retardants types, can be further separated into several different classes:

- Minerals such as aluminium hydroxide (ATH), magnesium hydroxide (MDH), huntite and hydromagnesite, various hydrates, red phosphorus, and boron compounds, mostly borates.

- Organohalogen compounds. This class includes organochlorines such as chlorendic acid derivatives and chlorinated paraffins; organobromines such as decabromodiphenyl ether (decaBDE), decabromodiphenyl ethane (a replacement for decaBDE), polymeric brominated compounds such as brominated polystyrenes, brominated carbonate oligomers (BCOs), brominated epoxy oligomers (BEOs), tetrabromophthalic anyhydride, tetrabromobisphenol A (TBBPA) and hexabromocyclododecane (HBCD). Most but not all halogenated flame retardants are used in conjunction with a synergist to enhance their efficiency. Antimony trioxide is widely used but other forms of antimony such as the pentoxide and sodium antimonate are also used.

- • Organophosphorus compounds. This class includes organophosphates such as triphenyl phosphate (TPP), resorcinol bis(diphenylphosphate) (RDP), bisphenol A diphenyl phosphate (BADP), and tricresyl phosphate (TCP); phosphonates such as dimethyl methylphosphonate (DMMP); and phosphinates such as aluminium diethyl phosphinate. In one important class of flame retardants, compounds contain both phosphorus and a halogen. Such compounds include tris(2,3-dibromopropyl) phosphate (brominated tris) and chlorinated organophosphates such as tris(1,3-dichloro-2-propyl)phosphate (chlorinated tris or TDCPP) and tetrakis(2-chlorethyl)dichloroisopentyldiphosphate (V6).

The mineral flame retardants mainly act as additive flame retardants and do not become chemically attached to the surrounding system. Most of the organohalogen and organophosphate compounds also do not react permanently to attach themselves into their surroundings but further work is now underway to graft further chemical groups onto these materials to enable them to become integrated without losing their retardant efficiency. This also will make these materials non emissive into the environment. Certain new non halogenated products, with these reactive and non emissive characteristics have been coming onto the market since 2010, because of the public debate about flame retardant emissions. Some of these new Reactive materials have even received US-EPA approval for their low environmental impacts.

Retardation Mechanisms

The basic mechanisms of flame retardancy vary depending on the specific flame retardant and the substrate. Additive and reactive flame-retardant chemicals can both function in the vapor (gaseous) or condensed (solid) phase.

Endothermic Degradation

Some compounds break down endothermically when subjected to high temperatures. Magnesium and aluminium hydroxides are an example, together with various carbonates and hydrates such as mixtures of huntite and hydromagnesite. The reaction removes heat from the substrate, thereby cooling the material. The use of hydroxides and hydrates is limited by their relatively low decomposition temperature, which limits the maximum processing temperature of the polymers (typically used in polyolefins for wire and cable applications).

Thermal Shielding (Solid Phase)

A way to stop spreading of the flame over the material is to create a thermal insulation barrier between the burning and unburned parts. Intumescent additives are often employed; their role is to turn the polymer surface into a char, which separates the flame from the material and slows the heat transfer to the unburned fuel. Non-halogenated inorganic and organic phosphate flame retardants typically act through this mechanism by generating a polymeric layer of charred phosphoric acid.

Dilution of Gas Phase

Inert gases (most often carbon dioxide and water) produced by thermal degradation of some materials act as diluents of the combustible gases, lowering their partial pressures and the partial pressure of oxygen, and slowing the reaction rate.

Gas Phase Radical Quenching

Chlorinated and brominated materials undergo thermal degradation and release hydrogen chloride and hydrogen bromide or, if used in the presence of a synergist like antimony trioxide, antimony halides. These react with the highly reactive H· and OH· radicals in the flame, resulting in an inactive molecule and a Cl· or Br· radical. The halogen radical is much less reactive compared to H· or OH·, and therefore has much lower potential to propagate the radical oxidation reactions of combustion.

Use and Effectiveness

Fire Safety Standards

Flame retardants are typically added to industrial and consumer products to meet flammability standards for furniture, textiles, electronics, and building products like insulation.

In 1975, California began implementing Technical Bulletin 117 (TB 117), which requires that materials such as polyurethane foam used to fill furniture be able to withstand a small open flame, equivalent to a candle, for at least 12 seconds. In polyurethane foam, furniture manufacturers typically meet TB 117 with additive halogenated organic flame retardants. Although no other U.S. states have a similar standard, because California has such a large market many manufacturers meet TB 117 in products that they distribute across the United States. The proliferation of flame retardants, and especially halogenated organic flame retardants, in furniture across the United States is strongly linked to TB 117.

In response to concerns about the health impacts of flame retardants in upholstered furniture, in February 2013 California proposed modifying TB 117 to require that fabric covering upholstered furniture meet a smolder test and to eliminate the foam flammability standards. Gov. Jerry Brown signed the modified TB117-2013 in November and it became effective in 2014. The modified regulation does not mandate a reduction in flame retardants.

However, these questions of eliminating emissions into the environment from flame retardants can be solved by using a new classification of highly efficient flame retardants, which do not contain halogen compounds, and which can also be keyed permanently into the chemical structure of the foams used in the furniture and bedding industries. The resulting foams have been certified to produce no flame retardant emissions. This new technology is based on entirely newly developed "Green Chemistry" with the final foam containing about one third by weight of natural oils. Use of this technology in the production of California TB 117 foams, would allow continued protection for the consumer against open flame ignition whilst providing the newly recognized and newly needed protection, against chemical emissions into home and office environments. More recent work during 2014 with this "Green Chemistry" has shown that foams containing about fifty percent of natural oils can be made which produce far less smoke when involved in fire situations. The ability of these low emission foams to reduce smoke emissions by up to 80% is an interesting property which will aid escape from fire situations and also lessen the risks for first responders i.e. emergency services in general and fire department personnel in particular.

In Europe, flame retardant standards for furnishings vary, and are their most stringent in the UK and Ireland. Generally the ranking of the various common flame retardant tests worldwide for furniture and soft furnishings would indicate that the California test Cal TB117 - 2013 test is the most straightforward to pass, there is increasing difficulty in passing Cal TB117 -1975 followed by the British test BS 5852 and followed by Cal TB133. One of the most demanding flammability tests worldwide is probably the US Federal Aviation Authority test for aircraft seating which involves the use of a kerosene burner which blasts flame at the test piece. The 2009 Greenstreet Berman study, carried out by the UK government, showed that in the period between 2002 and 2007 the UK Furniture and Furnishings Fire Safety Regulations accounted for 54 fewer deaths per year, 780 fewer non-fatal casualties per year and 1065 fewer fires each year following the introduction of the UK furniture safety regulations in 1988.

Effectiveness

The effectiveness of flame retardant chemicals at reducing the flammability of consumer products in house fires is disputed. Advocates for the flame retardant industry, such as the American Chemistry Council's North American Flame Retardant Alliance, cite a study from the National Bureau of Standards indicating that a room filled with flame-retarded products (a polyurethane foam-padded chair and several other objects, including cabinetry and electronics) offered a 15-fold greater time window for occupants to escape the room than a similar room free of flame retardants. However, critics of this position, including the lead study author, argue that the levels of flame retardant used in the 1988 study, while found commercially, are much higher than the levels required by TB 117 and used broadly in the United States in upholstered furniture.

Another study concluded flame retardants are an effective tool to reduce fire risks without creating toxic emissions.

Several studies in the 1980s tested ignition in whole pieces of furniture with different upholstery and filling types, including different flame retardant formulations. In particular, they looked at maximum heat release and time to maximum heat release, two key indicators of fire danger. These studies found that the type of fabric covering had a large influence on ease of ignition, that cotton fillings were much less flammable than polyurethane foam fillings, and that an interliner material substantially reduced the ease of ignition. They also found that although some flame retardant formulations decreased the ease of ignition, the most basic formulation that met TB 117 had very little effect. In one of the studies, foam fillings that met TB 117 had equivalent ignition times as the same foam fillings without flame retardants. A report from the Proceedings of the Polyurethane Foam Association also showed no benefit in open-flame and cigarette tests with foam cushions treated with flame retardants to meet TB 117. However, other scientists support this open-flame test.

Environmental and Health Issues

The environmental behaviour of flame retardants has been studied since the 1990s. Mainly brominated flame retardants were found in many environmental compartments and organisms including humans, and some individual substance were found to have toxic properties. Therefore, alternatives have been demanded by authorities, NGOs and equipment manufacturers. The EU-funded collaborative research project ENFIRO (EU research project FP7: 226563, concluded in 2012) started out from the assumption that not enough environmental and health data were known of alternatives to the established brominated flame retardants. In order to make the evaluation fully comprehensive, it was decided to compare also material and fire performance as well as attempt a life cycle assessment of a reference product containing halogen free versus brominated flame retardants. About a dozen halogen free flame retardants were studied representing a large variety of applications, from engineering plastics, printed circuit boards, encapsulants to textile and intumescent coatings. A large group of the studied flame retardants were found to have a good environmental and health profile: ammonium polyphosphate (APP), aluminium diethylphosphinate (Alpi), aluminium hydroxide (ATH), magnesium hydroxide (MDH), melamine polyphosphate (MPP), dihydrooxaphosphaphenanthrene (DOPO), zinc stannate (ZS) and zinc hydroxstannate (ZHS). Overall, they were found to have a much lower tendency to bioaccumulate in fatty tissue than the studied brominated flame retardants.

The tests on the fire behaviour of materials with different flame retardants revealed that halogen free flame retardants produce less smoke and toxic fire emissions, with the exception of the aryl phosphates RDP and BDP in styrenic polymers. The leaching experiments showed that the nature of the polymer is a dominating factor and that the leaching behaviour of halogen free and brominated flame retardants is comparable. The more porous or "hydrophilic" a polymers is the more flame retardants can be released. However, moulded plates which represent real world plastic products showed much lower leaching levels than extruded polymer granules. The impact assessment studies reconfirmed that the improper waste and recycling treatment of electronic products with brominated flame retardants can produce dioxins which is not the case with halogen free alternatives. Furthermore, the United States Environmental Protection Agency (US-EPA) has been carrying out a series of projects related to the environmental assessment of alternative flame retardants, the "design for environment" projects on flame retardants for printed wiring boards and alternatives to decabromo diphenylethers and hexabromocyclododecane (HBCD).

In 2009, the U.S. National Oceanic and Atmospheric Administration (NOAA) released a report on polybrominated diphenyl ethers (PBDEs) and found that, in contrast to earlier reports, they were found throughout the U.S. coastal zone. This nationwide survey found that New York's Hudson Raritan Estuary had the highest overall concentrations of PBDEs, both in sediments and shellfish. Individual sites with the highest PBDE measurements were found in shellfish taken from Anaheim Bay, California, and four sites in the Hudson Raritan Estuary. Watersheds that include the Southern California Bight, Puget Sound, the central and eastern Gulf of Mexico off the coast of Tampa and St. Petersburg, in Florida, and the waters of Lake Michigan near Chicago and Gary, Indiana, also were found to have high PBDE concentrations.

Health Concerns

The earliest flame retardants, polychlorinated biphenyls (PCBs), were banned in the U.S. in 1977 when it was discovered that they were toxic. Industries used brominated flame retardants instead, but these are now receiving closer scrutiny. In 2004 and 2008 the EU banned several types of polybrominated diphenyl ethers (PBDEs). Negotiations between the EPA and the two U.S. producers of DecaBDE (a flame retardant that has been used in electronics, wire and cable insulation, textiles, automobiles and airplanes, and other applications), Albemarle Corporation and Chemtura Corporation, and the largest U.S. importer, ICL Industrial Products, Inc., resulted in commitments by these companies to phase out decaBDE for most uses in the United States by December 31, 2012, and to end all uses by the end of 2013. The state of California has listed the flame retardant chemical chlorinated Tris (tris(1,3-dichloro-2-propyl) phosphate or TDCPP) as a chemical known to cause cancer. In December 2012, the California nonprofit Center for Environmental Health filed notices of intent to sue several leading retailers and producers of baby products for violating California law for failing to label products containing this cancer-causing flame retardant. While the demand for brominated and chlorinated flame retardants in North America and Western Europe is declining, it is rising in all other regions.

Nearly all Americans tested have trace levels of flame retardants in their body. Recent research links some of this exposure to dust on television sets, which may have been generated from the heating of the flame retardants in the TV. Careless disposal of TVs and other appliances such as microwaves or old computers may greatly increase the amount of environmental contamination. A recent study conducted by Harley *et al.* 2010 on pregnant women, living in a low-income, predominantly Mexican-immigrant community in California showed a significant decrease in fecundity associated with PBDE exposure in women.

Another study conducted by Chevrier *et al.* 2010 measured the concentration of 10 PBDE congeners, free thyroxine (T4), total T4, and thyroid-stimulating hormone (TSH) in 270 pregnant women around the 27th week of gestation. Associations between PBDEs and free and total T4 were found to be statistically insignificant. However, authors did find a significant association amongst exposure to PBDEs and lower TSH during pregnancy, which may have implications for maternal health and fetal development.

A prospective, longitudinal cohort study initiated after 11 September 2001, including 329 mothers who delivered in one of three hospitals in lower Manhattan, New York, was conducted by Herbstman *et al.* 2010. Authors of this study analyzed 210 cord blood specimens for selected PBDE congeners and assessed neurodevelopmental effects in the children at 12–48 and 72 months of age. Results showed that children who had higher cord blood concentrations of polybrominated diphenyl ethers (PBDEs) scored lower on tests of mental and motor development at 1–4 and 6 years of age. This was the first study to report any such associations in humans.

A similar study was conducted by Roze et al. 2009 in The Netherlands on 62 mothers and children to estimate associations between 12 Organohalogen compounds (OHCs), including polychlorinated biphenyls (PCBs) and brominated diphenyl ether (PBDE) flame retardants, measured in maternal serum during the 35th week of pregnancy and motor performance (coordination, fine motor skills), cognition (intelligence, visual perception, visuomotor integration, inhibitory control, verbal memory, and attention), and behavior scores at 5–6 years of age. Authors demonstrated for the first time that transplacental transfer of polybrominated flame retardants was associated with the development of children at school age.

Another study was conducted by Rose et al. in 2010 to measure circulating PBDE levels in 100 children between 2 and 5 years of age from California. The PBDE levels according to this study, in 2- to 5-year-old California children was 10 to 1,000 fold higher than European children, 5 times higher than other U.S. children and 2 to 10 times higher than U.S. adults. They also found that diet, indoor environment, and social factors influenced children's body burden levels. Eating poultry and pork contributed to elevated body burdens for nearly all types of flame retardants. Study also found that lower maternal education was independently and significantly associated with higher levels of most flame retardant congeners in the children.

San Antonio Statement on Brominated and Chlorinated Flame Retardants 2010: A group of 145 prominent scientists from 22 countries signed the first-ever consensus statement documenting health hazards from flame retardant chemicals found at high levels in home furniture, electronics, insulation, and other products. This statement documents that, with limited fire safety benefit, these flame retardants can cause serious health issues, and, as types of flame retardants are banned, the alternatives should be proven safe before being used. The group also wants to change widespread policies that require use of flame retardants.

A number of recent studies suggest that dietary intake is one of the main routes to human exposure to PBDEs. In recent years, PBDEs have become widespread environmental pollutants, while body burden in the general population has been increasing. The results do show notable coincidences between the China, Europe, Japan, and United States such as dairy products, fish, and seafood being a cause of human exposure to PBDEs due to the environmental pollutant.

A February 2012 study genetically engineered female mice to have mutations in the x-chromosome MECP2 gene, linked to Rett syndrome, a disorder in humans similar to autism. After exposure to BDE-47 (a PDBE) their offspring, who were also exposed, had lower birth weights and survivability and showed sociability and learning deficits.

A January 2013 study of mice showed brain damage from BDP-49, via inhibiting of the mitochondrial ATP production process necessary for brain cells to get energy. Toxicity was at very low levels. The study offers a possible pathway by which PDBEs lead to autism.

Mechanisms of Toxicity

Direct Exposure

Many halogenated flame retardants with aromatic rings, including most brominated flame retardants, are likely thyroid hormone disruptors. The thyroid hormones triiodothyronine (T3) and thyroxine (T4) carry iodine atoms, another halogen, and are structurally similar to many aromatic halogenated flame retardants, including PCBs, TBB-PA, and PBDEs. Such flame retardants therefore appear to compete for binding sites in the thyroid system, interfering with normal function of thyroid transport proteins (such as transthyretin) in vitro and thyroid hormone receptors. A 2009 *in vivo* animal study conducted by the US Environmental Protection Agency (EPA) demonstrated that deiodination, active transport, sulfation, and glucuronidation may be involved in disruption of thyroid homeostasis after perinatal exposure to PBDEs during critical developmental time points in utero and shortly after birth. Disruption of deiodinase as reported in the Szabo et al., 2009 *in vivo* study was supported in a follow-up *in vitro* study. The adverse effects on hepatic mechanism of thyroid hormone disruption during development have been shown to persist into adulthood. The EPA noted that PBDEs are particularly toxic to the developing brains of animals. Peer-reviewed studies have shown that even a single dose administered to mice during development of the brain can cause permanent changes in behavior, including hyperactivity.

Based on *in vitro* laboratory studies, several flame retardants, including PBDEs, TBB-PA, and BADP, likely also mimic other hormones, including estrogens, progesterone, and androgens. Bisphenol A compounds with lower degrees of bromination seem to exhibit greater estrogenicity. Some halogenated flame retardants, including the less-brominated PBDEs, can be direct neurotoxicants in *in vitro* cell culture studies: By altering calcium homeostasis and signalling in neurons, as well as neurotransmitter release and uptake at synapses, they interfere with normal neurotransmission. Mitochondria may be particularly vulnerable to PBDE toxicity due to their influence on oxidative stress and calcium activity in mitochondria. Exposure to PBDEs can also alter neural cell differentiation and migration during development.

Degradation Products

Many flame retardants degrade into compounds that are also toxic, and in some cases the degradation products may be the primary toxic agent:

- Halogenated compounds with aromatic rings can degrade into dioxins and dioxin-like compounds, particularly when heated, such as during production, a fire, recycling, or exposure to sun. Chlorinated dioxins are among the highly toxic compounds listed by the Stockholm Convention on Persistent Organic Pollutants.

- Polybrominated diphenyl ethers with higher numbers of bromine atoms, such as decaBDE, are less toxic than PBDEs with lower numbers of bromine atoms, such as pentaBDE. However, as the higher-order PBDEs degrade biotically or abiotically, bromine atoms are removed, resulting in more toxic PBDE congeners.

- When some halogenated flame retardants such as PBDEs are metabolized, they form hydroxylated metabolites that can be more toxic than the parent compound. These hydroxylated metabolites, for example, may compete more strongly to bind with transthyretin or other components of the thyroid system, can be more potent estrogen mimics than the parent compound, and can more strongly affect neurotransmitter receptor activity.

- Bisphenol-A diphenyl phosphate (BADP) and tetrabromobisphenol A (TBBPA) likely degrade to bisphenol A (BPA), an endocrine disruptor of concern.

Routes of Exposure

People can be exposed to flame retardants through several routes, including diet; consumer products in the home, vehicle, or workplace; occupation; or environmental contamination near their home or workplace. Residents in North America tend to have substantially higher body levels of flame retardants than people who live in many other developed areas, and around the world human body levels of flame retardants have increased over the last 30 years.

Exposure to PBDEs has been studied the most widely. As PBDEs have been phased out of use due to health concerns, organophosphorus flame retardants, including halogenated organophosphate flame retardants, have frequently been used to replace them. In some studies, indoor air concentrations of phosphorus flame retardants has been found to be greater than indoor air concentrations of PBDEs. The European Food Safety Authority (EFSA) issued in 2011 scientific opinions on the exposure to HBCD and TBBPA and its derivates in food and concluded that current dietary exposure in the European Union does not raise a health concern

Exposure in the General Population

The body burden of PBDEs in Americans correlates well with the level of PBDEs measured in swabs of their hands, likely picked up from dust. Dust exposure may occur in the home, car, or workplace. Levels of PBDEs can be as much as 20 times higher in vehicle dust as in household dust, and heating of the vehicle interior on hot summer days can break down flame retardants into more toxic degradation products. However, blood serum levels of PBDEs appear to correlate most highly with levels found in dust in the home. Perhaps 20% to 40% of adult U.S. exposure to PBDEs is through food intake, with the remaining exposure largely due to dust inhalation or ingestion.

Infants and toddlers are particularly exposed to halogenated flame retardants found in breast milk and dust. Because many halogenated flame retardants are fat-soluble, they accumulate in fatty areas such as breast tissue and are mobilized into breast milk, delivering high levels of flame retardants to breast-feeding infants. And, as consumer products age, small particles of material become dust particles in the air and land on surfaces around the home, including the floor. Young children crawling and playing on the floor frequently bring their hands to their mouths, ingesting about twice as much house dust as adults per day in the United States. Young children in the United States tend to carry higher levels of flame retardants per unit body weight than do adults.

Occupational Exposure

Some occupations expose workers to higher levels of halogenated flame retardants and their degradation products. A small study of U.S. foam recyclers and carpet installers, who handle padding often made from recycled polyurethane foam, showed elevated levels of flame retardants in their tissues. Workers in electronics recycling plants around the world also have elevated body levels of flame retardants relative to the general population. Environmental controls can substantially reduce this exposure, whereas workers in areas with little oversight can take in very high levels of flame retardants. Electronics recyclers in Guiyu, China, have some of the highest human body levels of PBDEs in the world. A study conducted in Finland determined the occupational exposure of workers to brominated flame retardants and chlorinated flame retardants (TBBPA, PBDEs, DBDPE, HBCD, Hexabromobenzene and Dechlorane plus). In 4 recycling sites of waste electrical and electronic equipment (WEEE), the study concluded that control measures implemented on site significantly reduced the exposure. Workers making products that contain flame retardants (such as vehicles, electronics, and baby products) may be similarly exposed. U.S. firefighters can have elevated levels of PBDEs and high levels of brominated furans, toxic degradation products of brominated flame retardants.

Environmental Exposure

Flame retardants manufactured for use in consumer products have been released into environments around the world. The flame retardant industry has developed a voluntary initiative to reduce emissions to the environment (VECAP) by promoting best practices during the manufacturing process. Communities near electronics factories and disposal facilities, especially areas with little environmental oversight or control, develop high levels of flame retardants in air, soil, water, vegetation, and people.

Organophosphorus flame retardants have been detected in wastewater in Spain and Sweden, and some compounds do not appear to be removed thoroughly during water treatment.

Disposal

When products with flame retardants reach the end of their usable life, they are typically recycled, incinerated, or landfilled.

Recycling can contaminate workers and communities near recycling plants, as well as new materials, with halogenated flame retardants and their breakdown products. Electronic waste, vehicles, and other products are often melted to recycle their metal components, and such heating can generate toxic dioxins and furans. When wearing Personal Protection Equipment (PPE) and when a ventilation system is installed, exposure of workers to dust can be significantly reduced, as shown in the work conducted by the recycling plant Stena-Technoworld AB in Sweden. Brominated flame retardants may also change the physical properties of plastics, resulting in inferior performance in recycled products and in "downcycling" of the materials. It appears that plastics with brominated flame retardants are mingling with flame-retardant-free plastics in the recycling stream and such downcycling is taking place.

Poor-quality incineration similarly generates and releases high quantities of toxic degradation products. Controlled incineration of materials with halogenated flame retardants, while costly, substantially reduces release of toxic byproducts.

Many products containing halogenated flame retardants are sent to landfills. Additive, as opposed to reactive, flame retardants are not chemically bonded to the base material and leach out more easily. Brominated flame retardants, including PBDEs, have been observed leaching out of landfills in industrial countries, including Canada and South Africa. Some landfill designs allow for leachate capture, which would need to be treated. These designs also degrade with time.

Opposition

The widespread use of flame retardants in the United States evolved after California enacted Technical Bulletin 117 (TB117) in 1975 requiring fillings in furniture such as polyurethane foam to resist an open flame for 12 seconds. In 2013, a Chicago Tribune investigative series alleged that the chemical and tobacco industries mounted a campaign to increase the amount of flame retardants in homes while avoiding the need to manufacture a fire safe cigarette. US Senators asked the EPA to evaluate flame retardants for possible health risks. Firefighters concerned about high cancer rates in their profession have called for stricter regulation of use of flame retardants in homes.

California's furniture flammability standards were changed in 2014. TB117-2013 allows manufacturers to market products that withstand a smolder test in lieu of the open flame test. There are legislative attempts to ban or restrict the use of certain flame retardants.

TBB

Flame retardants are effective in reducing the flammability of synthetic materials. The EPA has conducted an assessment of new flame retardants, such as 2,3,4,5-tetrabromo-ethyl-hexylbenzoate (TBB). However, long-term toxicological investigations into the cumulative effects of chronic TBB exposure were not done as they were outside the scope of the review.

TB117

California Technical Bulletin 117 was developed by the California Bureau of Home Furnishings through a consensus standards development process and first implemented in 1975. This regulation was intended to prevent ignition or slow the spread of the flame if the furniture is the first to ignite. When fires do occur, multiple studies show that foams treated with flame retardants burn much slower than untreated foam, giving occupants time to escape.

National Bureau of Standards Testing

In a 1988 test program was conducted by the former National Bureau of Standards (NBS), now the National Institute of Standards and Technology (NIST), to quantify the effects of fire retardant chemicals on total fire hazard. Five different types of products, each made from a different type of plastic were used. The products were made up in analogous fire-retardant (FR) and non-retarded variants (NFR).

The impact of FR (flame retardant) materials on the survivability of the building occupants was assessed in two ways:

First, comparing the time until a domestic space is not fit for occupation in the burning room, known as "untenability"; this is applicable to the occupants of the burning room. Second, comparing the total production of heat, toxic gases, and smoke from the fire; this is applicable to occupants of the building remote from the room of fire origin.

The time to untenability is judged by the time that is available to the occupants before either (a) room flashover occurs, or (b) untenability due to toxic gas production occurs. For the FR tests, the average available escape time was more than 15-fold greater than for the occupants of the room without fire retardants.

Hence, with regard to the production of combustion products,

- The amount of material consumed in the fire for the fire retardant (FR) tests was less than half the amount lost in the non-fire retardant (NFR) tests.

- The FR tests indicated an amount of heat released from the fire which was 1/4 that released by the NFR tests.

- The total quantities of toxic gases produced in the room fire tests, expressed in "CO equivalents," were 1/3 for the FR products, compared to the NFR ones.

- The production of smoke was not significantly different between the room fire tests using NFR products and those with FR products.

Thus, in these tests, the fire retardant additives decreased the overall fire hazard.

Global Demand

In 2013, the world consumption of flame retardants was more than 2 million tonnes. The commercially most import application area is the construction sector. It needs flame retardants for instance for pipes and cables made of plastics. In 2008 the United States, Europe and Asia consumed 1.8 million tonnes, worth US$4.20-4.25 billion. According to Ceresana, the market for flame retardants is increasing due to rising safety standards worldwide and the increased use of flame retardants. It is expected that the global flame retardant market will generate US$5.8 billion. In 2010, Asia-Pacific was the largest market for flame retardants, accounting for approximately 41% of global demand, followed by North America, and Western Europe.

Polychlorinated Biphenyl

A polychlorinated biphenyl (PCB) is an organic chlorine compound with the formula $C_{12}H_{10-x}Cl_x$. Polychlorinated biphenyls were once widely deployed as dielectric and coolant fluids in electrical apparatus, carbonless copy paper and in heat transfer fluids. Because of their longevity, PCBs are still widely in use, even though their manufacture has declined drastically since the 1960s, when a host of problems were identified. Because of PCBs' environmental toxicity and classification as a persistent organic pollutant, PCB production was banned by the United States Congress in 1979 and by the Stockholm Convention on Persistent Organic Pollutants in 2001. The International Research Agency on Cancer (IRAC), rendered PCBs as definite carcinogens in humans. According to the U.S. Environmental Protection Agency (EPA), PCBs cause cancer in animals and are probable human carcinogens. Many rivers and buildings including schools, parks, and other sites are contaminated with PCBs, and there have been contaminations of food supplies with the toxins.

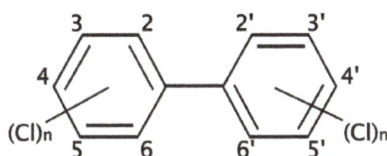

Chemical structure of PCBs. The possible positions of chlorine atoms on the benzene rings are denoted by numbers assigned to the carbon atoms.

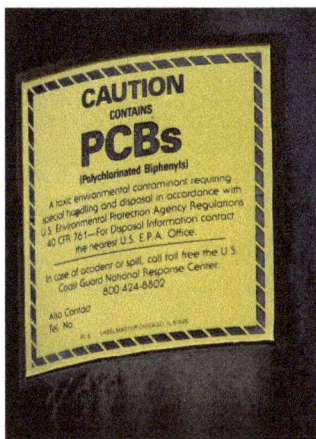

PCB warning label on a power transformer known to contain PCBs.

Some PCBs share a structural similarity and toxic mode of action with dioxin. Other toxic effects such as endocrine disruption (notably blocking of thyroid system functioning) and neurotoxicity are known. The maximum allowable contaminant level in drinking water in the United States is set at zero, but because of water treatment technologies, a level of 0.5 parts per billion is the de facto level.

The bromine analogues of PCBs are polybrominated biphenyls (PBBs), which have analogous applications and environmental concerns.

Physical and Chemical Properties

Physical Properties

The compounds pale-yellow viscous liquids. They are hydrophobic, with low water solubilities — 0.0027-0.42 ng/L for Aroclors, but they have high solubilities in most organic solvents, oils, and fats. They have low vapor pressures at room temperature.

They have dielectric constants of 2.5~2.7, very high thermal conductivity, high flash points (from 170 to 380 °C)

The density varies from 1.182 to 1.566 kg/L. Other physical and chemical properties vary widely across the class. As the degree of chlorination increases, melting point and lipophilicity increase, and vapour pressure and water solubility decrease.

PCBs do not easily break down or degrade, which made them attractive to industry. PCB mixtures are resistant to acids, bases, oxidation, hydrolysis and temperature change.

They can generate extremely toxic dibenzodioxins and dibenzofurans through partial oxidation. Intentional degradation as a treatment of unwanted PCBs generally requires high heat or catalysis.

PCBs readily penetrate skin, PVC (polyvinyl chloride), and latex (natural rubber). PCB-resistant materials include Viton, polyethylene, polyvinyl acetate (PVA), polytetrafluoroethylene (PTFE), butyl rubber, nitrile rubber, and Neoprene.

Structure and Toxicity

PCB's are derived from biphenyl, which has the formula $C_{12}H_{10}$, sometimes written $(C_6H_5)_2$. In PCB's most of the H's are replaced by chloride. It is a mixture of compounds, given the single identifying CAS number 1336-36-3). There are 209 configurations with 1 to 10 chlorine atoms, of which about 130 are found in commercial PCBs.

Toxic effects vary depending on the specific PCB. In terms of their structure and toxicity, PCBs fall into 2 distinct categories, referred to as coplanar or non-*ortho*-substituted arene substitution patterns and noncoplanar or *ortho*-substituted congeners.

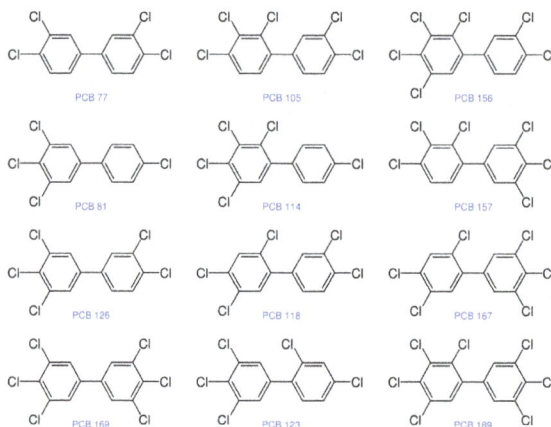

Structures of the twelve dioxin-like PCBs

Coplanar or Non-Ortho

The coplanar group members have a fairly rigid structure, with their two phenyl rings in the same plane. It renders their structure similar to polychlorinated dibenzo-p-dioxins (PCDDs) and polychlorinated dibenzofurans, and allows them to act like PCDDs, as an agonist of the aryl hydrocarbon receptor (AhR) in organisms. They are considered as contributors to overall dioxin toxicity, and the term dioxins and dioxin-like compounds is often used interchangeably when the environmental and toxic impact of these compounds is considered.

Noncoplanar

Noncoplanar PCBs, with chlorine atoms at the *ortho* positions cause neurotoxic and immunotoxic effects, but at levels much higher than normally associated with dioxins. They do not activate the AhR, and are not considered part of the dioxin group, and as of 1998 had been of less concern to regulatory bodies.

Di-ortho-substituted, non-coplanar PCBs interfere with intracellular signal transduction dependent on calcium which may lead to neurotoxicity. In 2000 it was shown that ortho-PCBs can disrupt thyroid hormone transport by binding to transthyretin.

Alternative Names

Commercial PCB mixtures were marketed under the following names:

Brazil	Japan	United States
• Ascarel	• Kanechlor (used by Kanegafuchi)	• Aroclor xxxx (used by Monsanto Company)
Former Czechoslovakia	• Santotherm (used by Mitsubishi)	• Asbestol
• Delor	• Pyroclor	• Askarel
France	**Former USSR**	• Bakola131
• Phenoclor	• Sovol	• Chlorextol - Allis-Chalmers trade name
• Pyralène (both used by Prodolec)	• Sovtol	• Hydol
Germany	**United Kingdom**	• Inerteen (used by Westinghouse)
• Clophen (used by Bayer)	• Aroclor xxxx (used by Monsanto Company)	• Noflamol
Italy	• Askarel	• Pyranol/Pyrenol, Chlorinol (widely used in GE's oil-filled "chlorinol"-branded metal can capacitors, detected by a pungent characteristic odor released by them (especially when they fail) in appliances/consumer & commercial electronic units & motors that utilized them from the early 1960s-late 1970s such as many A/C units, Seeburg Jukeboxes & Zenith TVs) (used by General Electric)
• Apirolio		• Saf-T-Kuhl
• Fenclor		• Therminol FR Series (Monsanto ceased production in 1971).

Aroclor Mixtures

The only North American producer, Monsanto Company, marketed PCBs under the trade name Aroclor from 1930 to 1977. These were sold under trade names followed by a 4-digit number. In general, the first two digits refer to the number of carbon atoms in the biphenyl skeleton (for PCBs this is 12); the second two numbers indicate the percentage of chlorine by mass in the mixture. Thus, Aroclor 1260 has 12 carbon atoms and contains 60% chlorine by mass. An exception is Aroclor 1016, which also has 12 carbon atoms, but has 42% chlorine by mass. Aroclor 1016 was prepared by the fractional distillation of Aroclor 1242, which excluded the higher boiling (i.e., more highly chlorinated) congeners.

Different Aroclors were used at different times and for different applications. In electrical equipment manufacturing in the USA, Aroclor 1260 and Aroclor 1254 were the main mixtures used before 1950; Aroclor 1242 was the main mixture used in the 1950s and 1960s until it was phased out in 1971 and replaced by Aroclor 1016.

Production

One estimate (2006) suggested that 1 M tons of PCB's had been produced. 40% of this material was thought to remain in use. Another estimate put the total global production of PCBs on the order of 1.5 million tons. The United States was the single largest producer with over 600,000 tons produced between 1930 and 1977. The European region follows with nearly 450,000 tons through 1984. It is unlikely that a full inventory of global PCB production will ever be accurately tallied, as there were factories in Poland, East Germany, and Austria that produced unknown amounts of PCBs.

Applications

PCB's utility was based largely on their chemical stability, including low flammability, and high dielectric constant. In an electric arc, PCB's generate incombustible gases. Use of PCBs is commonly divided into closed and open applications. Examples of closed applications include coolants and insulating fluids (transformer oil) for transformers and capacitors, such as those used in old fluorescent light ballasts, hydraulic fluids, lubricating and cutting oils, etc. In contrast, the major open application of PCBs was in carbonless copy ("NCR") paper, which even nowadays results in paper contamination. Other open applications were as plasticizers in paints and cements, stabilizing additives in flexible PVC coatings of electrical cables and electronic components, pesticide extenders, reactive flame retardants and sealants for caulking, adhesives, wood floor finishes, such as *Fabulon* and other products of Halowax in the U.S., de-dusting agents, waterproofing compounds, casting agents. Because of its use as a plasticizer in paints and especially "coal tars" that were used widely to coat water tanks, bridges and other infrastructure pieces.

Environmental Transport and Transformations

PCBs have entered the environment through both use and disposal. The environmental fate of PCBs is complex and global in scale.

Water

Because of their low vapour pressure, PCBs accumulate primarily in the hydrosphere, despite their hydrophobicity, in the organic fraction of soil, and in organisms.

The hydrosphere is the main reservoir. The immense volume of water in the oceans is still capable of dissolving a significant quantity of PCBs.

Air

A small volume of PCBs has been detected throughout the earth's atmosphere. The atmosphere serves as the primary route for global transport of PCBs, particularly for those congeners with one to four chlorine atoms. In the atmosphere, PCBs may be degraded by hydroxyl radicals, or directly by photolysis of carbon-chlorine bonds (even if this is a less important process).

Atmospheric concentrations of PCBs tend to be lowest in rural areas, where they are typically in the picogram per cubic meter range, higher in suburban and urban areas, and highest in city centres, where they can reach 1 ng/m^3 or more. In Milwaukee, an atmospheric concentration of 1.9 ng/m^3 has been measured, and this source alone was estimated to account for 120 kg/year of PCBs entering Lake Michigan. In 2008, concentrations as high as 35 ng/m^3, 10 times higher than the EPA guideline limit of 3.4 ng/m^3, have been documented inside some houses in the U.S.

Volatilization of PCBs in soil was thought to be the primary source of PCBs in the atmosphere, but research suggests ventilation of PCB-contaminated indoor air from buildings is the primary source of PCB contamination in the atmosphere.

Biosphere

In biosphere, PCBs can be degraded by either bacteria or eukaryotes, but the speed of the reaction depends on both the number and the disposition of chlorine atoms in the molecule: less substituted, meta- or para- substituted PCBs undergo biodegradation faster than more substituted congeners.

In bacteria, PCBs may be dechlorinated through reductive dechlorination, or oxidized by dioxygenase enzyme.

In eukaryotes, PCBs may be oxidized by the cytochrome P450 enzyme.

Like many lipiphilic toxins, PCBs biomagnify up the food chain. For instance, ducks can accumulate PCBs from eating fish and other aquatic life from contaminated rivers, and these can cause harm to human health or even death when eaten.

PCBs can be transported by birds from aquatic sources onto land via feces and carcasses.

Health Effects

The toxicity of PCBs varies considerably among congeners. The coplanar PCBs, known as nonortho PCBs because they are not substituted at the ring positions ortho to (next to) the other ring, (i.e. PCBs 77, 126, 169, etc.), tend to have dioxin-like properties, and generally are among the most toxic congeners. Because PCBs are almost invariably found in complex mixtures, the concept of toxic equivalency factors (TEFs) has been developed to facilitate risk assessment and regulation, where more toxic PCB congeners are assigned higher TEF values on a scale from 0 to 1. One of the most toxic compounds known, 2,3,7,8-tetrachlorodibenzo[p]dioxin, a PCDD, is assigned a TEF of 1.

Labelling transformers containing PCBs

Exposure and Excretion

In general individuals are exposed to PCBs overwhelmingly through food, much less so by breathing contaminated air, and least by skin contact. Once exposed, some PCBs may change to other chemicals inside the body. These chemicals or unchanged PCBs can be excreted in feces or may remain in a person's body for years, with half lives estimated at 10–15 years. PCBs collect in body fat and milk fat. PCB's biomagnify up the food web and are present in fish and waterfowl of contaminated aquifers. Infants are exposed to PCBs through breast milk or by intrauterine exposure through transplacental transfer of PCBs and are at the top of the food chain.

Signs and Symptoms

Humans

The most commonly observed health effects in people exposed to extremely high levels of PCBs are skin conditions, such as chloracne and rashes, but these were known to be symptoms of acute systemic poisoning dating back to 1922. Studies in workers exposed to PCBs have shown changes in blood and urine that may indicate liver damage. In Japan in 1968, 280 kg of PCB-contaminated rice bran oil was used as chicken feed, resulting in a mass poisoning, known as Yushō disease, in over 1800 people. Common symptoms included dermal and ocular lesions, irregular menstrual cycles and lowered immune responses. Other symptoms included fatigue, headaches, coughs, and unusual skin sores. Additionally, in children, there were reports of poor cognitive development.

Women exposed to PCBs before or during pregnancy can give birth to children with lowered cognitive ability, immune compromise, and motor control problems.

There is evidence that crash dieters that have been exposed to PCBs have an elevated risk of health complications. Stored PCBs in the adipose tissue becomes mobilized into the blood when individuals begin to crash diet. PCBs have shown toxic and mutagenic effects by interfering with hormones in the body. PCBs, depending on the specific congener, have been shown to both inhibit and imitate estradiol, the main sex hormone in females. Imitation of the estrogen compound can feed estrogen-dependent breast cancer cells, and possibly cause other cancers, such as uterine or cervical. Inhibition of estradiol can lead to serious developmental problems for both males and females, including sexual, skeletal, and mental development issues.

High PCB levels in adults have been shown to result in reduced levels of the thyroid hormone triiodothyronine, which affects almost every physiological process in the body, including growth and development, metabolism, body temperature, and heart rate. It also resulted in reduced immunity and increased thyroid disorders.

Animals

Animals that eat PCB-contaminated food even for short periods of time suffer liver damage and may die. In 1968 in Japan, 400,000 birds died after eating poultry feed that was contaminated with PCBs. Animals that ingest smaller amounts of PCBs in food over several weeks or months develop various health effects, including anemia; acne-like skin conditions (chloracne); liver, stomach, and thyroid gland injuries (including hepatocarcinoma),, and thymocyte apoptosis. Other effects of PCBs in animals include changes in the immune system, behavioral alterations, and impaired reproduction. PCBs that have dioxin-like activity are known to cause a variety of teratogenic effects in animals. Exposure to PCBs causes hearing loss and symptoms similar to hypothyroidism in rats.

Cancer

In 2013, the International Agency for Research on Cancer (IARC) classified dioxin-like PCBs as human carcinogens. According to the U.S. EPA, PCBs have been shown to cause cancer in animals and evidence supports a cancer-causing effect in humans. Per EPA, studies have found increases in malignant melanoma and rare liver cancers in PCB workers.

In 2013, the International Association for Research on Cancer (IARC) determined that the evidence for PCBs causing non-Hodgkin Lymphoma is "limited" and "not consistent". In contrast an association between elevated blood levels of PCBs and non-Hodgkin lymphoma had been previously accepted.

PCBs may play a role in the development of cancers of the immune system because some tests of laboratory animals subjected to very high doses of PCBs have shown effects on the animals' immune system, and some studies of human populations have reported an association between environmental levels of PCBs and immune response.

History

In 1865 the first "PCB-like" chemical was discovered, and was found to be a byproduct of coal tar. Years later in 1881, German chemists synthesized the first PCB in a laboratory. Between then and 1914, large amounts of PCBs were released into the environment, to the extent that there are still measurable amounts of PCBs in feathers of birds currently held in museums.

Old power transformers are a major legacy source of PCBs. Even units not originally filled with PCB may be contaminated, since PCB and oil mix freely and any given transformer may have been refilled from hoses or tanks also used with PCBs.

In 1935, Monsanto Chemical Company (now Solutia Inc) took over commercial production of PCBs from Swann Chemical Company which had begun in 1929. PCBs, originally termed "chlorinated diphenyls", were commercially produced as mixtures of isomers at different degrees of chlorination. The electric industry used PCBs as a non-flammable replacement for mineral oil to cool and insulate industrial transformers and capacitors. PCBs were also commonly used as heat stabilizer in cables and electronic components to enhance the heat and fire resistance of PVC.

In the 1930s, the toxicity associated with PCBs and other chlorinated hydrocarbons, including polychlorinated naphthalenes, was recognized because of a variety of industrial incidents. Between 1936 and 1937, there were several medical cases and papers released on the possible link between PCBs and its detrimental health effects. In 1936 a U.S. Public Health Service official described the wife and child of a worker from the Monsanto Industrial Chemical Company who exhibited blackheads and pustules on their skin. The official attributed these symptoms to contact with the worker's clothing after he returned from work. In 1937, a conference about the hazards was organized at Harvard School of Public Health, and a number of publications referring to the toxicity of various chlorinated hydrocarbons were published before 1940. In 1947 Robert Brown reminded chemists that Arochlors were "objectionably toxic. Thus the maximum permissible concentration for an 8-hr. day is 1 mg/m³ of air. They also produce a serious and disfiguring dermatitis".

In 1954 Japan, Kanegafuchi Chemical Co. Ltd. (Kaneka Corporation) first produced PCB's, and continued until 1972.

Through the 1960s Monsanto Chemical Company knew increasingly more about PCB's harmful effects on humans and the environment, per internal leaked documents released in 2002, yet PCB manufacture and use continued with few restraints until the 1970s.

In 1966, PCBs were determined by Swedish chemist Dr. Soren Jensen to be an environmental contaminant. Jensen, according to a 1994 article in Sierra, named chemicals PCBs, which previously, had simply been called "phenols" or referred to by various trade names, such as Aroclor, Kanechlor, Pyrenol, Chlorinol and others.

In 1972, PCB production plants existed in Austria, the then Federal Republic of Germany, France, Great Britain, Italy, Japan, Spain, USSR, and USA.

In the early 1970s, Ward B. Stone of the New York State Department of Environmental Conservation (NYSDEC) first published his findings that PCBs were leaking from transformers and had contaminated the soil at the bottom of utility poles.

There have been allegations that Industrial Bio-Test Laboratories engaged in data falsification in testing relating to PCBs.

In 2003, Monsanto and Solutia Inc., a Monsanto corporate spin-off, reached a $700 million settlement with the residents of West Anniston, Alabama who had been affected by the manufacturing and dumping of PCBs. In a trial lasting six weeks, the jury found that "Monsanto had engaged in outrageous behavior, and held the corporations and its corporate successors liable on all six counts it considered - including negligence, nuisance, wantonness and suppression of the truth."

Existing products containing PCBs which are "totally enclosed uses" such as insulating fluids in transformers and capacitors, vacuum pump fluids, and hydraulic fluid, are allowed to remain in use.

The public, legal, and scientific concerns about PCBs arose from research indicating they are likely carcinogens having the potential to adversely impact the environment and, therefore, undesirable as commercial products. Despite active research spanning five decades, extensive regulatory actions, and an effective ban on their production since the 1970s, PCBs still persist in the environment and remain a focus of attention.

Pollution Due to PCBs

Belgium

In 1999, the Dioxin Affair occurred when 50 kg of PCB transformer oils were added to a stock of recycled fat used for the production of 500 tonnes of animal feed, eventually affecting around 2,500 farms in several countries. The name *Dioxin Affair* was coined from early misdiagnosis of dioxins as the primary contaminants, when in fact they turned out to be a relatively small part of the contamination caused by thermal reactions of PCBs. The PCB congener pattern suggested the contamination was from a mixture of Aroclor 1260 & 1254. Over 9 million chickens, and 60,000 pigs were destroyed because of the contamination. The extent of human health effects has been debated, in part because of the use of differing risk assessment methods. One group predicted increased cancer rates, and increased rates of neurological problems in those exposed as neonates. A second study suggested carcinogenic effects were unlikely and that the primary risk would be associated with developmental effects due to exposure in pregnancy and neonates. Two businessmen who knowingly sold the contaminated feed ingredient received two-year suspended sentences for their role in the crisis.

Italy

The Italian company Caffaro, located in Brescia, specialized in producing PCBs from 1938 to 1984, following the acquisition of the exclusive rights to use the patent in Italy from Monsanto. The pollution resulting from this factory and the case of Anniston, in the USA, are the largest known cases in the world of PCB contamination in water and soil, in terms of the amount of toxic substance dispersed, size of the area contaminated, number of people involved and duration of production.

The values reported by the local health authority (ASL) of Brescia since 1999 are 5,000 times above the limits set by Ministerial Decree 471/1999 (levels for residential areas, 0.001 mg/kg). As a result of this and other investigations, in June 2001, a complaint of an environmental disaster was presented to the Public Prosecutor's Office of Brescia. Research on the adult population of Brescia showed that residents of some urban areas, former workers of the plant, and consumers of contaminated food, have PCB levels in their bodies that are in many cases 10-20 times higher than reference values in comparable general populations. PCBs entered the human food supply by animals grazing on contaminated pastures near the factory, especially in local veal mostly eaten by farmers' families. The exposed population showed an elevated risk of Non-Hodgkin lymphoma, but not for other specific cancers.

Japan

In 1968, a mixture of dioxins and PCBs got into rice bran oil produced in northern Kyushu. Contaminated cooking oil sickened more than 1860 people. The symptoms were called Yushō disease.

In Okinawa, high levels of PCB contamination in soil on Kadena Air Base were reported in 1987 at thousands of parts per million, some of the highest levels found in any pollution site in the world.

Republic of Ireland

In December 2008, a number of Irish news sources reported testing had revealed "extremely high" levels of dioxins, by toxic equivalent, in pork products, ranging from 80 to 200 times the EU's upper safe limit of 1.5 pg WHO-TEQ$_{DFP}$/μg i.e. 0.12 to 0.3 parts per billion.

Brendan Smith, the Minister for Agriculture, Fisheries and Food, stated the pork contamination was caused by PCB-contaminated feed that was used on 9 of Ireland's 400 pig farms, and only one feed supplier was involved. Smith added that 38 beef farms also used the same contaminated feed, but those farms were quickly isolated and no contaminated beef entered the food chain. While the contamination was limited to just 9 pig farms, the Irish government requested the immediate withdrawal and disposal of all pork-containing products produced in Ireland and purchased since 1 September 2008. This request for withdrawal of pork products was confirmed in a press release by the Food Safety Authority of Ireland on December 6.

It is thought that the incident resulted from the contamination of fuel oil used in a drying burner at a single feed processor, with PCBs. The resulting combustion produced a highly toxic mixture of PCBs, dioxins and furans, which was included in the feed produced and subsequently fed to a large number of pigs.

Kenya

In Kenya, a number of cases have been reported in the 2010s of thieves selling transformer oil, stolen from electric transformers, to the operators of roadside food stalls for use in deep frying. When used for frying, it is reported that transformer oil lasts much longer than regular cooking oil. The downside of this misuse of the transformer oil is the threat to the health of the consumers, due to the presence of PCBs.

Slovakia

The chemical plant Chemko in Strážske (east Slovakia) was an important producer of polychlorinated biphenyls for the former communist block (Comecon) until 1984. Chemko contaminated a large part of east Slovakia, especially the sediments of the Laborec river and reservoir Zemplínska šírava.

United Kingdom

Monsanto manufactured PCBs at its chemical plant in Newport, South Wales, until the mid- to late-1970s. During this period, waste matter, including PCBs, from the Newport site was dumped at a disused quarry near Groes-faen, west of Cardiff, from where it continues to be released in waste water discharges.

Spain

Several cetacean species have very high mean blubber PCB concentrations likely to cause population declines and suppress population recovery. Striped dolphins, bottle-nose dolphins and killer whales were found to have mean levels that markedly exceeded all known marine mammal PCB toxicity thresholds. The western Mediterranean Sea and the south-west Iberian Peninsula were identified as "hotspots".

United States

Alabama

PCBs (manufactured through most of the 20th century) originating from Monsanto Chemical Company in Anniston, Alabama were dumped into Snow Creek, which then spread to Choccolocco Creek, then Logan Martin Lake. In the early 2000s, class action lawsuits were settled by local land owners, including those on Logan Martin Lake, and Lay Reservoir (downstream on the Coosa River), for the PCB pollution. Donald Stewart, former Senator from Alabama, first learned of the concerns of hundreds of west Anniston residents after representing a church which had been approached about selling its property by Monsanto. Stewart went on the be the pioneer and lead attorney in the first and majority of cases against Monsanto and focused on residents in the immediate area known to be most polluted. Other attorneys later joined in to file suits for those outside the main immediate area around the plant; one of these was the late Johnnie Cochran.

In 2007, the highest pollution levels remained concentrated in Snow and Choccolocco Creeks. Concentrations in fish have declined and continue to decline over time; sediment disturbance, however, can resuspend the PCBs from the sediment back into the water column and food web.

Great Lakes

In 1976 environmentalists found PCBs in the sludge at Waukegan Harbor, the south-west end of Lake Michigan. They were able to trace the source of the PCBs back to the Outboard Marine Corporation that was producing boat motors next to the harbor. By 1982, the Outboard Marine Corporation was court-ordered to release quantitative data referring to their PCB waste released. The data stated that from 1954 they released 100,000 tons of PCB into the environment, and that the sludge contained PCBs in concentrations as high as 50%.

Late during the construction of new on- and off-ramps in the M-13 interchange on the Zilwaukee bridge approach, workers uncovered an uncharted landfill containing PCB-contaminated waste, necessitating an environmental cleanup. In August 22, 1989,

The Detroit Free Press noted that the clean up costs would cost over $100,000 and delay the opening of the ramps to the M-13 interchange in Zilwaukee, which were scheduled for opening that year.

Much of the Great Lakes area were still heavily polluted with PCBs in 1988, despite extensive remediation work. Locally caught fresh water fish and shellfish are contaminated with PCBs, and their consumption is restricted.

Indiana

From the late 1950s through 1977, Westinghouse Electric used PCBs in the manufacture of capacitors in its Bloomington, Indiana plant. Reject capacitors were hauled and dumped in area salvage yards and landfills, including Bennett's Dump, Neal's Landfill and Lemon Lane Landfill. Workers also dumped PCB oil down factory drains, which contaminated the city sewage treatment plant. The City of Bloomington gave away the sludge to area farmers and gardeners, creating anywhere from 200 to 2000 sites, which remain unaddressed. Over 2 million pounds of PCBs were estimated to have been dumped in Monroe and Owen counties. Although federal and state authorities have been working on the sites' environmental remediation, many areas remain contaminated. Concerns have been raised regarding the removal of PCBs from the karst limestone topography, and regarding the possible disposal options. To date, the Westinghouse Bloomington PCB Superfund site case does not have a Remedial Investigation/Feasibility Study (RI/FS) and Record of Decision (ROD), although Westinghouse signed a US Department of Justice Consent Decree in 1985. The 1985 consent decree required Westinghouse to construct an incinerator that would incinerate PCB-contaminated materials. Because of public opposition to the incinerator, however, the State of Indiana passed a number of laws that delayed and blocked its construction. The parties to the consent decree began to explore alternative remedies in 1994 for six of the main PCB contaminated sites in the consent decree. Hundreds of sites remain unaddressed as of 2014. Monroe County will never be PCB-free, as noted in a 2014 Indiana University program about the local contamination.

On 15 February 2008, Monroe County approved a plan to clean up the three remaining contaminated sites in the City of Bloomington, at a cost of $9.6 million to CBS Corp., the successor of Westinghouse. In 1999, Viacom bought CBS, so they are current responsible party for the PCB sites.

Massachusetts

Pittsfield, in western Massachusetts, was home to the General Electric (GE) transformer and capacitor divisions, and electrical generating equipment built and repaired in Pittsfield powered the electrical utility grid throughout the nation. PCB-contaminated oil routinely migrated from GE's 254-acre (1.03 km²) industrial plant located in the very center of the city to the surrounding groundwater, nearby Silver Lake, and to the

Housatonic River, which flows through Massachusetts, Connecticut, and down to Long Island Sound. PCB-containing solid material was widely used as fill, including oxbows of the Housatonic River. Fish and waterfowl who live in and around the river contain significant levels of PCBs and are not safe to eat.

New Bedford Harbor, which is a listed Superfund site, contains some of the highest sediment concentrations in the marine environment.

New York

Pollution of the Hudson River is largely due to dumping of PCBs by General Electric from 1947 to 1977. GE dumped an estimated 1.3 million pounds of PCBs into the Hudson River during these years. This pollution caused a range of harmful effects to wildlife and people who eat fish from the river or drink the water.

Love Canal is a neighborhood in Niagara Falls, New York that was heavily contaminated with toxic waste including PCBs.

Eighteen Mile Creek in Lockport, New York is an EPA Superfund site for PCBs contamination.

North Carolina

One of the largest deliberate PCB spills in American history occurred in the summer of 1978 when 31,000 gallons of PCB-contaminated oil were illegally sprayed in 3-foot (0.91 m) swaths along the roadsides of some 240 miles (390 km) of North Carolina highway shoulders in 14 counties and at the Fort Bragg Army Base. The crime, known as "the midnight dumpings", occurred over nearly 2 weeks, as drivers of a black-painted tanker truck drove down one side of rural Piedmont highways spraying PCB-laden waste and then up the other side the following night.

Under Governor James B. Hunt, Jr., state officials then erected large, yellow warning signs along the contaminated highways that read: "CAUTION: PCB Chemical Spills Along Highway Shoulders." The illegal dumping is believed to have been motivated by the passing of the Toxic Substances Control Act (TSCA), which became effective on August 2, 1978 and increased the expense of chemical waste disposal.

Within a couple of weeks of the crime, Robert Burns and his sons, Timothy and Randall, were arrested for dumping the PCBs along the roadsides. Burns was a business partner of Robert "Buck" Ward, Jr., of the Ward PCB Transformer Company, in Raleigh. Burns and sons pleaded guilty to state and Federal criminal charages; Burns received a three to five-year prison sentence. Ward was acquitted of state charges in the dumping, but was sentenced to 18 months prison time for violation of TSCA.

Cleanup and disposal of the roadside PCBs generated controversy, as the Governor's plan to pick up the roadside PCBs and to bury them in a landfill in rural Warren County were strongly opposed in 1982 by local residents.

In October 2013, at the request of the South Carolina Department of Health and Environmental Control (SCDHEC), the City of Charlotte, North Carolina decided to stop applying sewage sludge to land while authorities investigated the source of PCB contamination. In February 2014, the City of Charlotte admitted PCBs have entered their sewage treatment centers as well.

After the 2013 SCDHEC had issued emergency regulations the City of Charlotte discovered high levels of PCB's entering its sewage waste water treatment plants, where sewage is converted to sewage sludge. The city at first denied it had a problem, then admitted an "event" occurred in February 2014, and in April that the problem had occurred much earlier. The city stated that its very first test with a newly changed test method revealed very high PCB levels in its sewage sludge farm field fertilizer. Because of the widespread use of the contaminated sludge, SCDHEC subsequently issued PCB fish advisories for nearly all streams and rivers bordering farm fields that had been applied with city waste.

Ohio

The Clyde cancer cluster (also known as the Sandusky County cancer cluster) is a childhood cancer cluster that has affected many families in Clyde, Ohio and surrounding areas. PCBs were found in soil in a public park within the area of the cancer cluster.

In Akron, Ohio, soil was contaminated and noxious PCB-laden fumes had been put into the air by an electrical transformer deconstruction operation from the 1930s to the 1960s.

South Carolina

From 1955 until 1977, the Sangamo Weston plant in Pickens, SC, used PCBs to manufacture capacitors, and dumped 400,000 pounds of PCB contaminated wastewater into the Twelve Mile Creek. In 1990, the EPA declared the 228 acres (0.92 km^2) site of the capacitor plant, its landfills and the polluted watershed, which stretches nearly 1,000 acres (4.0 km^2) downstream to Lake Hartwell as a Superfund site. Two dams on the Twelve Mile Creek are to be removed and on Feb. 22, 2011 the first of two dams began to be dismantled. Some contaminated sediment is being removed from the site and hauled away, while other sediment is pumped into a series of settling ponds.

In 2009, the state environmental regulators SCDHEC noted fish species in Lake Wateree contained exceptionally high levels of PCB contamination and posted adviseries that fish from the lake were unsafe to eat.

In 2013, the state environmental regulators issued a rare emergency order, banning all sewage sludge from being land applied or deposited on landfills, as it contained very high levels of PCBs. The problem had not been discovered until thousands of acres of farm land in the state had been contaminated by the hazardous sludge. A criminal investigation to determine the perpetrator of this crime was launched.

Washington

As of 2015, several bodies of water in the state of Washington were contaminated with PCBs, including the Columbia River, the Duwamish River, Green Lake, Lake Washington, the Okanogan River, Puget Sound, the Spokane River, the Walla Walla River, the Wenatchee River, and the Yakima River. A study by Washington State published in 2011 found that the two largest sources of PCB flow into the Spokane River were City of Spokane stormwater (44%), municipal and industrial discharges (20%). PCBs entered the environment through paint, hydraulic fluids, sealants, inks and have been found in river sediment and wild life. Spokane utilities will spend $300 million to prevent PCBs from entering the river in anticipation of a 2017 federal deadline to do so. In August 2015 Spokane joined other U.S cities like San Diego and San Jose, California, and Westport, Massachusetts. in seeking damages from Monsanto.

Wisconsin

From 1954 until 1971, the Fox River in Appleton, Wisconsin had PCBs deposited into it from Appleton Paper/NCR, P.H. Gladfelter, Georgia Pacific and other notable local paper manufacturing facilities. The Wisconsin DNR estimates that after wastewater treatment the PCB discharges to the Fox River due to production losses ranged from 81,000 kg to 138,000 kg. (178,572 lbs. to 304,235 lbs). The production of Carbon Copy Paper and its byproducts led to the discharge into the river. Fox River clean up is ongoing.

Regulation

In 1972 the Japanese government banned the production, use, and import of PCBs.

In 1973, the use of PCBs in "open" or "dissipative" sources, such as plasticisers in paints and cements, casting agents, fire retardant fabric treatments and heat stabilizing additives for PVC electrical insulation, adhesives, paints and waterproofing, railroad ties was banned in Sweden.

In 1979, concern over the toxicity and persistence (chemical stability) of PCBs in the environment led the United States Congress to ban their domestic production.

In 1981, the UK banned closed uses of PCBs in new equipment, and nearly all UK PCB synthesis ceased; closed uses in existing equipment containing in excess of 5 litres of PCBs were not stopped until December 2000.

Methods of Destruction

Physical

PCBs are technically attractive because of their inertness, which includes their resistance to combustion. Nonetheless, they can be effectively destroyed by Incineration at 1000 °C. When combusted at lower temperatures, they convert in part to more hazardous materials, including dibenzofurans and dibenzodioxins. When conducted properly, the combustion products are water, carbon dioxide, and hydrogen chloride. In some cases, the PCB's are combusted as a solution in kerosene. PCB's have also been destroyed by pyrolysis in the presence of alkali metal carbonates.

Chemical

PCBs are fairly chemically unreactive, this property being attractive for its application as an inert material. They resist oxidation. Many chemical compounds are available to destroy or reduce the PCBs. Commonly, PCBs are degraded by basis mixtures of glycols, which displace some or all chloride. Also effective are reductants such as sodium or sodium naphthenide.

Microbial

Some micro-organisms degrade PCBs by reducing the C-Cl bonds. Microbial dechlorination tends to be rather slow-acting in comparison to other methods, on PCB as a soil contaminant. Many microbes that work well in laboratory conditions, do not function in the field. Enzymes extracted from microbes can show PCB activity. Especially promising seems to be the use of vitamin B12. In 2005, *Shewanella oneidensis* biodegraded a high percentage of PCBs in soil samples.

Homologs

For a complete list of the 209 PCB congeners, see PCB congener list. Note that biphenyl, while not technically a PCB congener because of its lack of chlorine substituents, is still typically included in the literature.

PCB homolog	CASRN	Cl substituents	Number of congeners
Biphenyl (not a PCB)	92-52-4	0	1
Monochlorobiphenyl	27323-18-8	1	3
Dichlorobiphenyl	25512-42-9	2	12
Trichlorobiphenyl	25323-68-6	3	24

Tetrachlorobiphenyl	26914-33-0	4	42
Pentachlorobiphenyl	25429-29-2	5	46
Hexachlorobiphenyl	26601-64-9	6	42
Heptachlorobiphenyl	28655-71-2	7	24
Octachlorobiphenyl	55722-26-4	8	12
Nonachlorobiphenyl	53742-07-7	9	3
Decachlorobiphenyl	2051-24-3	10	1

Polybrominated Diphenyl Ethers

Polybrominated diphenyl ethers or PBDEs, are organobromine compounds that are used as flame retardant. Like other brominated flame retardants, PBDEs have been used in a wide array of products, including building materials, electronics, furnishings, motor vehicles, airplanes, plastics, polyurethane foams, and textiles. They are structurally akin to the PCBs and other polyhalogenated compounds, consisting of two halogenated aromatic rings. PBDEs are classified according to the average number of bromine atoms in the molecule. The health hazards of these chemicals have attracted increasing scrutiny, and they have been shown to reduce fertility in humans at levels found in households. Their chlorine analogs are polychlorinated diphenyl ethers (PCDEs). Because of their toxicity and persistence, the industrial production of some PBDEs is restricted under the Stockholm Convention, a treaty to control and phase out major persistent organic pollutants (POPs).

Classes of PBDEs

The family of PBDEs consists of 209 possible substances, which are called congeners (PBDE = $C_{12}H_{(10-x)}Br_xO$ (x = 1, 2, ..., 10 = m + n)). The number of isomers for mono-, di-, tri-, tetra-, penta-, hexa-, hepta-, octa-, nona-, and decabromodiphenyl ethers are 3, 12, 24, 42, 46, 42, 24, 12, 3 and 1, respectively. In the United States, PBDEs are marketed with trade names: DE-60F, DE-61, DE-62, and DE-71 applied to pentaBDE mixtures; DE-79 applied to octaBDE mixtures; DE 83R and Saytex 102E applied to decaBDE mixtures. The available commercial PBDE products are not single compounds or even single congeners but rather a mixture of congeners.

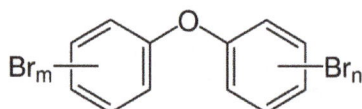

Chemical structure of PBDEs

Lower Brominated PBDEs

These species average 1-5 bromine atoms per molecule and are regarded as more dangerous because they more efficiently bioaccumulate. Lower-brominated PBDEs have been known to affect hormone levels in the thyroid gland. Studies have linked them to reproductive and neurological risks at certain concentrations or higher.

Higher Brominated PBDEs

These species average more than 5 bromine atoms per molecule.

The commercial mixture, named pentabromodiphenyl ether, contains the pentabromo derivative predominantly (50-62%), however the mixture also contains tetrabromides (24-38%) and hexabromides (4-8%), as well as traces of the tribromides (0-1%). In similar manner, commercial octabromodiphenyl ether is a mixture of homologs: hexa-, hepta-, octa-, nona-, and decabromides.

Health and Environmental Concerns

Since the 1990s, environmental concerns were raised because of the high lipophilicity of PBDEs and their high resistance to degradation processes. While biodegradation is not considered the main pathway for PBDEs, the photolysis and pyrolysis can be of interest in studies of transformation of PBDEs. People are exposed to low levels of PBDEs through ingestion of food and by inhalation. PBDEs bioaccumulate in blood, breast milk, and fat tissues. Personnel associated with the manufacture of PBDE-containing products are exposed to highest levels of PBDEs. Bioaccumulation is of particular concern in such instances, especially for personnel in recycling and repair plants of PBDE-containing products. People are also exposed to these chemicals in their domestic environment because of their prevalence in common household items. Studies in Canada have found significant concentrations of PBDEs in common foods such as salmon, ground beef, butter, and cheese. PBDEs have also been found at high levels in indoor dust, sewage sludge, and effluents from wastewater treatment plants. Increasing PBDE levels have been detected in the blood of marine mammals such as harbor seals.

There is also growing concern that PBDEs share the environmental long life and bioaccumulation properties of polychlorinated dibenzodioxins.

Case Studies

A non-peer-reviewed study of 20 mother-child pairs in the United States conducted by the Environmental Working Group found that the median blood levels of PBDEs in children (62 parts per billion) were 3.2 times higher than in their mothers. PBDEs have also been shown to have hormone-disrupting effects, in particular, on estrogen and thyroid hormones. A 2009 animal study conducted by the US Environmental Protection Agency (EPA) demonstrates that deiodination, active transport, sulfation, and glucuronidation may be involved in disruption of thyroid homeostasis after perinatal exposure to PBDEs during critical developmental time points in utero and shortly after birth. The adverse effects on hepatic mechanism of thyroid hormone disruption during development have been shown to persist into adulthood. The EPA noted that PBDEs are particularly toxic to the developing brains of animals. Peer-reviewed studies have shown that even a single dose administered to mice during development of the brain can cause permanent changes in behavior, including hyperactivity.

Swedish scientists first reported substances related to PentaBDE were accumulating in human breast milk. Studies by the Swedish Society for Nature Conservation found for the first time very high levels of higher brominated PBDEs (BDE-209) in eggs of Peregrine falcons. Two forms of PBDEs, Penta and Octa, are no longer manufactured in the United States because of health and safety concerns. Based on a comprehensive risk assessment under the Existing Substances Regulation 793/93/EEC, the EU has completely banned the use of Penta and Octa BDE since 2004. However, both chemicals are still found in furniture and foam items made before the phase-out was completed. The most-common PBDEs that are used in electronics are in a form known as Deca. Deca is banned in Europe for this use and in some U.S. states. For PBDE, EPA has set reference dose of 7 micrograms per kilogram of body weight, which is "believed to be without appreciable effects". However, Linda Birnbaum, PhD, a senior toxicologist formerly with the EPA (now NIEHS) notes concern: "What I see is another piece of evidence that supports the fact that levels of these chemicals in children appear to be higher than the levels in their parents; I think this study raises a red flag." Previous study by EWG in 2003 published test results showing that the average level of fire-retardants in breast milk from 20 American mothers was 75 times higher than the average levels measured in Europe.

It has been postulated that increasing levels of PBDEs in the environment could be correlated with the increasing incidence of feline hyperthyroidism. However, a study in 2007 found that no association could be detected between hyperthyroid cats and serum PBDE levels.

An experiment conducted at Woods Hole Oceanographic Institution in 2005 showed that the isotopic signature of methoxy-PBDEs found in whale blubber contained carbon-14, the naturally occurring radioactive isotope of carbon. Methoxy-PBDEs are produced by some marine species. If the methoxy-PBDEs in the whale had come from artificial (human-made) sources, they would have contained only carbon-12 and no carbon-14 due to the fact that virtually all PBDEs that are produced artificially use petroleum as the source of carbon; all carbon-14 would have long since completely decayed from that source. The isotopic signatures of the PBDEs themselves were not evaluated. The carbon-14 may instead be in methoxy groups enzymatically added to man-made PBDEs.

A 2010 study found that children with higher concentrations of PBDE congeners 47, 99 and 100 in their umbilical cord blood at birth scored lower on tests of mental and physical development between the ages of one and six. Developmental effects were particularly evident at four years of age, when verbal and full IQ scores were reduced 5.5 to 8.0 points for those with the highest prenatal exposures.

Regulations of PBDEs

As of June 1, 2006 the State of California began prohibiting the manufacture, distribution, and processing of flame-retardant products containing pentabrominated diphenyl ether (pentaBDE) and octabrominated diphenyl (octaBDE). PBDEs are ubiquitous in the environment, and, according to the EPA, exposure may pose health risks. According to U.S. EPA's Integrated Risk Information System, evidence indicates that PBDEs may possess liver toxicity, thyroid toxicity, and neurodevelopmental toxicity. In June 2008, the U.S. EPA set a safe daily exposure level ranging from 0.1 to 7 ug per kg body weight per day for 4 most common PBDEs. In April 2007, the legislature of the state of Washington passed a bill banning the use of PBDEs. The State of Maine Department of Environmental Protection found that all PBDEs should be banned. In August, 2003, the State of California outlawed the sale of penta- and octa- PBDE and products containing them, effective January 1, 2008. In May 2007, the legislature of the state of Maine passed a bill phasing out the use of DecaBDE.

The European Union decided to ban the use of two classes of flame retardants, in particular, polybrominated diphenyl ethers (PBDEs) and polybrominated biphenyls (PBBs) in electric and electronic devices. This ban was formalised in the RoHS Directive, and an upper limit of 1 g/kg for the sum of PBBs and PBDEs was set. In February 2009, the Institute for Reference Materials and Measurements (IRMM) released two certified reference materials (CRMs) to help analytical laboratories better detect these two classes of flame retardants. The reference materials were custom-made to contain all relevant PBDEs and PBBs at levels close to the legal limit.

At an international level, in May 2009 the Parties of the Stockholm Convention for Persistent Organic Pollutants (POPs) took the decision to list commercial penta-BDE and commercial octa-BDE as POP substances. This listing is due to the properties of hexa-BDE (hexabromodiphenyl ether) and hepta-BDE (heptabromodiphenyl ether) which are the main components of commercial octa-BDE, and due to the properties of tetra-BDE (tetrabromodiphenyl ether) and penta-BDE (pentabromodiphenyl ether) which are the main components of commercial penta-BDE.

Hexabromocyclododecane

Hexabromocyclododecane (HBCD or HBCDD) is a brominated flame retardant. It consists of twelve carbon, eighteen hydrogen, and six bromine atoms tied to the ring. Its primary application is in extruded (XPS) and expanded (EPS) polystyrene foam that is used as thermal insulation in the building industry. Other uses are upholstered furniture, automobile interior textiles, car cushions and insulation blocks in trucks, packaging material, video cassette recorder housing and electric and electronic equipment.

According to UNEP, "HBCD is produced in China, Europe, Japan, and the USA. The known current annual production is approximately 28,000 tonnes per year. The main share of the market volume is used in Europe and China" (figures from 2009/2010).

HBCD's toxicity and its harm to the environment are currently discussed. HBCD can be found in environmental samples such as birds, mammals, fish and other aquatic organisms as well as soil and sediment. On this basis, on 28 October 2008 the European Chemicals Agency decided to include HBCD in the SVHC list, Substances of Very High Concern, within the Registration, Evaluation, Authorisation and Restriction of Chemicals framework. On 18 February 2011, HBCD was listed in the Annex XIV of REACH and hence is subject to Authorisation. HBCD can be used until the so-called "sunset date" (21 August 2015). After that date, only authorised applications will be allowed in the EU. HBCD has been found widely present in biological samples from remote areas and supporting evidences for its classification as Persistent, Bioaccumulative and Toxic (PBT) and undergoes long-range environmental transportation. In July 2012, an EU harmonised classification and labelling for HBCD entered into force. HBCD has been classified as a category 2 for reproductive toxicity. Since August 2010 hexabromocyclododecanes are included in the EPA's List of Chemicals of Concern. On May 2013 the Stockholm Convention on Persistent Organic Pollutants (POPs) decided to include HBCD in the Convention's Annex A for elimination, with specific exemptions for expanded and extruded polystyrene in buildings needed to give countries time to phase-in safer substitutes. HBCD is listed for elimination, but with a specific exemption for expanded polystyrene (EPS) and extruded polystyrene (XPS) in buildings. Countries may choose to use this exemption for up to five years after the request for exemption is submitted. Japan was the first country to implement a ban on the import and production of HBCD effective in May 2014.

Because HBCD has 16 possible stereo-isomers with different biological activities, the substance poses a difficult problem for manufacture and regulation. The HBCD commercial mixture is composed of three main diastereomers denoted as alpha (α-HBCD), beta (β-HBCD) and gamma (γ-HBCD) with traces of others. A series of four published in vivo mice studies were conducted between several federal and academic institutions to characterize the toxicokinetic profiles of individual HBCD stereoisomers. The predominant diastereomer in the HBCD mixture, γ-HBCD, undergoes rapid hepatic metabolism, fecal and urinary elimination, and biological conversion to other diastereomers with a short biological half-life of 1–4 days. After oral exposure to the γ-HBCD diastereomer, β-HBCD was detected in the liver and brain, and α-HBCD and β-HBCD was detected in the fat and feces with multiple novel metabolites identified - monohydroxy-pentabromocyclododecane, monohydroxy-pentabromocyclododecene, dihydroxy-pentabromocyclododecene, and dihydroxy-pentabromocyclododecadiene. In contrast, α-HBCD is more biologically persistent, resistant to metabolism, bioaccumulates in lipid-rich tissues after a 10-day repeated exposure study, and has a longer biological half-life of up to 21 days; only α-HBCD was detected in the liver, brain, fat and feces with no stereoisomerization to γ-HBCD or β-HBCD and low trace levels of four different hydroxylated metabolites were identified. Developing mice had higher HBCD tissue levels than adult mice after exposure to either α-HBCD or γ-HBCD indicating the potential for increased susceptibility of the developing young to HBCD effects. The reported toxicokinetic differences of individual HBCD diastereoisomers have important implications for the extrapolation of toxicological studies of the commercial HBCD mixture to the assessment of human risk.

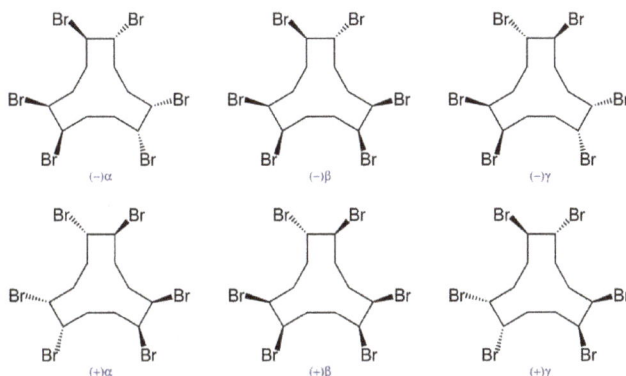

Structures of the six (out of 16 possible) hexabromocyclododecane isomers that are present in the technical product at > 1 %

Environmental Concerns

Due to its persistence, toxicity, and ecotoxicity, the Stockholm Convention on Persistent Organic Pollutants decided in May 2013 to list hexabromocyclododecane in Annex A to the Convention with specific exemptions for production and use in expanded polystyrene and extruded polystyrene in buildings. Countries may choose to use this exemption for up to five years after the request for exemption is submitted.

There is a large and still increasing stock of HBCD in the anthroposphere, mainly in EPS and XPS insulation boards. A long term environmental monitoring programme run by the Fraunhofer Institute for Molecular Biology and Applied Ecology demonstrates a general trend that HBCD concentrations are decreasing over time. HBCD emissions into the environment are controlled under the voluntary industry emission management programme: the Voluntary Emissions Control Action Programme (VECAP). The VECAP annual report demonstrates continuous decrease of potential emissions of HBCD to the environment.

Brominated Flame Retardant

Brominated flame retardants (BFRs) are organobromine compounds that have an inhibitory effect on combustion chemistry and tend to reduce the flammability of products containing them. Of all the commercialized chemical flame retardants, the brominated variety are widely used (19.7% of the market). They are effective in plastics and textile applications, e.g. electronics, clothes and furniture.

Types of Compounds

Many different BFRs are produced synthetically with widely varying chemical properties. There are several groups:

- Polybrominated diphenyl ethers (PBDEs): DecaBDE, OctaBDE (not manufactured anymore), PentaBDE (not manufactured anymore, the first BFR, commercialized in the 1950s)

- Polybrominated biphenyl (PBB), not manufactured anymore

- Brominated cyclohydrocarbons

- Other brominated flame retardants with different properties and mechanisms

Decabromodiphenyl ether (*Deca-BDE* or *DeBDE*) - In August 2012, the UK authorities proposed decabromodiphenyl ether (Deca-BDE or DeBDE) as a candidate for Authorisation under the EU's regulatory regime on chemicals, REACH. On 5 July 2013 ECHA withdrew Deca-BDE from its list of priority substances for Authorisation under REACH, therefore closing the public consultation. On 1 August 2014, ECHA submitted a restriction proposal for Deca-BDE. The agency is proposing a restriction on the manufacture, use and placing on the market of the substance and of mixtures and articles containing it. On 17 September 2014, ECHA submitted the restriction report which initiates a six months public consultation. A decision could be adopted by mid-2016.

Hexabromocyclododecane (*HBCD* or *HBCDD*) is a ring consisting of twelve carbon atoms with six bromine atoms tied to the ring. The commercially used HBCD is in fact a mixture of different isomers. HBCD is toxic to water-living organisms. The UNEP Stockholm Convention has listed HBCD for elimination, but allowing a temporary exemption for the use in polystyrene insulation foams in buildings.

Tetrabromobisphenol A (*TBBPA* or *TBBP-A*) is regarded as toxic to water environment. This flame retardant is mainly used in printed circuit boards, as a reactive. Since TBBPA is chemically bound to the resin of the printed circuit board, it is less easily released than the loosely applied mixtures in foams such that an EU risk assessment concluded in 2005 that TBBPA poses no risk to human health in that application. TBBPA is also used as an additive in acrylonitrile butadiene styrene (ABS).

Contents in Plastics

Content of brominated flame retardants in different polymers:

Polymer	Content [%]	Substances
Polystyrene foam	0.8–4	HBCD
High impact polystyrene	11–15	DecaBDE, brominated polystyrene

Epoxy resin	0–0.1	TBBPA
Polyamides	13–16	DecaBDE, brominated polystyrene
Polyolefins	5–8	DecaBDE, propylene dibromo styrene
Polyurethanes	n/a	No brominated FR available
Polyterephthalate	8–11	Brominated polystyrene
Unsaturated polyesters	13–28	TBBPA
Polycarbonate	4–6	Brominated polystyrene
Styrene copolymers	12–15	Brominated polystyrene

Production

390,000 tons of brominated flame retardants were sold in 2011. This represents 19.7% of the flame retardants market.

Types of Applications

The electronics industry accounts for the greatest consumption of BFRs. In computers, BFRs are used in four main applications: in printed circuit boards, in components such as connectors, in plastic covers, and in cables. BFRs are also used in a multitude of products, including, but not exclusively, plastic covers of television sets, carpets, pillows, paints, upholstery, and domestic kitchen appliances.

Testing for BFR in Plastics

Until recently testing for BFR has been cumbersome. Cycle time, cost and level of expertise required for the test engineer has precluded the implementation of any screening of plastic components in a manufacturing or in a product qualification/validation environment.

Recently, with the introduction of a new analytical instrument IA-Mass, screening of plastic material alongside a manufacturing line became possible. A five-minute detection cycle and a 20-minute quantification cycle is available to test and to qualify plastic parts as they reach the assembly line. IA-Mass identifies the presence of bromine (PBB, PBDE, and some others), but cannot characterize all the BFRs present in the plastic matrix.

In February 2009, the Institute for Reference Materials and Measurements (IRMM) released two certified reference materials (CRMs) to help analytical laboratories better detect two classes of flame retardants, namely polybrominated diphenyl ethers (PBDEs) and polybrominated biphenyls (PBBs). The two reference materials were custom made to contain all relevant PBDEs and PBBs at levels close to the legal limit set out in the RoHS Directive of 1 g/kg for the sum of PBBs and PBDEs.

Environmental and Safety Issues

Many brominated chemicals are under increasing criticism in their use in household furnishings and where children would come into contact with them. Some believe PBDEs could have harmful effects on humans and animals. Increasing concern has prompted some European countries to ban some of them, following the precautionary principle more common in Europe. Some PBDEs are lipophilic and bioaccumulative. PBDEs have been found in people all over the world.

Some brominated flame retardants were identified as persistent, bioaccumulative, and toxic to both humans and the environment and were suspected of causing neurobehavioral effects and endocrine disruption. One particular target group is Firefighters who are exposed to brominated fire retardants during firefighting operations and is resulting in cancer rates that far exceed the general public. As an example, in Europe, brominated flame retardants have gone through REACH and when risks were identified appropriate risk management options were put in place; such was the case for commercial Penta-BDE and commercial Octa-BDE. Given the current state of waste disposal in the world, there is a potential for BFRs to be released into the environment.

Organochloride

An organochloride, organochlorine compound, chlorocarbon, or chlorinated hydrocarbon is an organic compound containing at least one covalently bonded atom of chlorine that has an effect on the chemical behavior of the molecule. The chloroalkane class (alkanes with one or more hydrogens substituted by chlorine) provides common examples. The wide structural variety and divergent chemical properties of organochlorides lead to a broad range of names and applications. Organochlorides are very useful compounds in many applications, but some are of profound environmental concern.

Physical and Chemical Properties

Chlorination modifies the physical properties of hydrocarbons in several ways. The compounds are typically denser than water due to the higher atomic weight of chlorine versus hydrogen. Aliphatic organochlorides are alkylating agents because chloride is a leaving group.

Natural Occurrence

Many organochlorine compounds have been isolated from natural sources ranging from bacteria to humans. Chlorinated organic compounds are found in nearly every class of biomolecules including alkaloids, terpenes, amino acids, flavonoids, steroids, and fatty acids. Organochlorides, including dioxins, are produced in the high temperature environment of forest fires, and dioxins have been found in the preserved ashes of lightning-ignited fires that predate synthetic dioxins. In addition, a variety of simple chlorinated hydrocarbons including dichloromethane, chloroform, and carbon tetrachloride have been isolated from marine algae. A majority of the chloromethane in the environment is produced naturally by biological decomposition, forest fires, and volcanoes.

The natural organochloride epibatidine, an alkaloid isolated from tree frogs, has potent analgesic effects and has stimulated research into new pain medication. (The frogs obtain epibatidine through their diet and then sequester it on their skin. Likely dietary sources are beetles, ants, mites, and flies.)

Preparation

From Chlorine

Alkanes and aryl alkanes may be chlorinated under free radical conditions, with UV light. However, the extent of chlorination is difficult to control. Aryl chlorides may be prepared by the Friedel-Crafts halogenation, using chlorine and a Lewis acid catalyst.

The haloform reaction, using chlorine and sodium hydroxide, is also able to generate alkyl halides from methyl ketones, and related compounds. Chloroform was formerly produced thus.

Chlorine adds to the multiple bonds on alkenes and alkynes as well, giving di- or tetrachloro compounds.

Reaction with Hydrogen Chloride

Alkenes react with hydrogen chloride (HCl) to give alkyl chlorides. For example, the industrial production of chloroethane proceeds by the reaction of ethylene with HCl:

$$H_2C=CH_2 + HCl \rightarrow CH_3CH_2Cl$$

In oxychlorination, hydrogen chloride instead of the more expensive chlorine for the same purpose:

$$CH_2=CH_2 + 2\ HCl + \tfrac{1}{2}\ O_2 \rightarrow ClCH_2CH_2Cl + H_2O.$$

Secondary and tertiary alcohols react with hydrogen chloride to give the corresponding chlorides. In the laboratory, the related reaction involving zinc chloride in concentrat-ed hydrochloric acid:

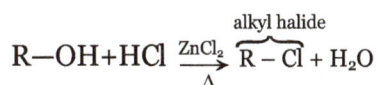

$$R-OH + HCl \xrightarrow[\Delta]{ZnCl_2} \overbrace{R-Cl}^{\text{alkyl halide}} + H_2O$$

Called the Lucas reagent, this mixture was once used in qualitatitve organic analysis for classifying alcohols.

Other Chlorinating Agents

Alkyl chlorides are most easily prepared by treating alcohols with thionyl chloride ($SOCl_2$) or phosphorus pentachloride (PCl_5), but also commonly with sulfuryl chloride (SO_2Cl_2) and phosphorus trichloride (PCl_3):

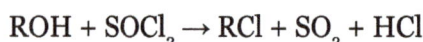

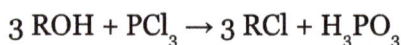

$$ROH + SOCl_2 \rightarrow RCl + SO_2 + HCl$$

$$3\ ROH + PCl_3 \rightarrow 3\ RCl + H_3PO_3$$

$$ROH + PCl_5 \rightarrow RCl + POCl_3 + HCl$$

In the laboratory, thionyl chloride is especially convenient, because the byproducts are gaseous. Alternatively, the Appel reaction can be used:

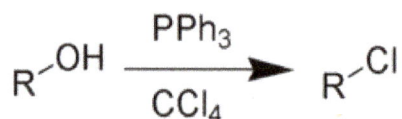

$$R\text{-}OH \xrightarrow[\text{CCl}_4]{\text{PPh}_3} R\text{-}Cl$$

Reactions

Alkyl chlorides are versatile building blocks in organic chemistry. While alkyl bromides and iodides are more reactive, alkyl chlorides tend to be less expensive and more readily available. Alkyl chlorides readily undergo attack by nucleophiles.

Heating alkyl halides with sodium hydroxide or water gives alcohols. Reaction with alkoxides or aroxides give ethers in the Williamson ether synthesis; reaction with thiols give thioethers. Alkyl chlorides readily react with amines to give substituted amines. Alkyl chlorides are substituted by softer halides such as the iodide in the Finkelstein reaction. Reaction with other pseudohalides such as azide, cyanide, and thiocyanate are possible as well. In the presence of a strong base, alkyl chlorides undergo dehydrohalogenation to give alkenes or alkynes.

Alkyl chlorides react with magnesium to give Grignard reagents, transforming an electrophilic compound into a nucleophilic compound. The Wurtz reaction reductively couples two alkyl halides to couple with sodium.

Applications

Vinyl Chloride

The largest application of organochlorine chemistry is the production of vinyl chloride. The annual production in 1985 was around 13 billion kilograms, almost all of which was converted into polyvinylchloride (PVC).

Chloromethanes

Most low molecular weight chlorinated hydrocarbons such as chloroform, dichloro-methane, dichloroethene, and trichloroethane are useful solvents. These solvents tend to be relatively non-polar; they are therefore immiscible with water and effective in cleaning applications such as degreasing and dry cleaning. Several billion kilograms of chlorinated methanes are produced annually, mainly by chlorination of methane:

$$CH_4 + x\ Cl_2 \rightarrow CH_{4-x}Cl_x + x\ HCl$$

The most important is dichloromethane, which is mainly used as a solvent. Chloro-methane is a precursor to chlorosilanes and silicones. Historically significant, but smaller in scale is chloroform, mainly a precursor to chlorodifluoromethane ($CHClF_2$) and tetrafluoroethene which is used in the manufacture of Teflon.

Pesticides

The two main groups of organochlorine insecticides are the DDT-type compounds and the chlorinated alicyclics. Their mechanism of action differs slightly: The DDT like compounds work on the peripheral nervous system. At the axon's sodium channel, they prevent gate closure after activation and membrane depolarization. Sodium ions leak through the nerve membrane and create a destabilizing negative "afterpotential" with hyperexcitability of the nerve. This leakage causes repeated discharges in the neuron either spontaneously or after a single stimulus.

Chlorinated cyclodienes include aldrin, dieldrin, endrin, heptachlor, chlordane and en-dosulfan. A 2- to 8-hour exposure leads to depressed central nervous system (CNS) activity, followed by hyperexcitability, tremors, and then seizures. The mechanism of action is the insecticide binding at the $GABA_A$ site in the gamma-Aminobutyric acid (GABA) chloride ionophore complex, which inhibits chloride flow into the nerve.

Other examples include dicofol, mirex, kepone and pentachlorophenol. These can be either hydrophilic or hydrophobic depending on their molecular structure.

Insulators

Polychlorinated biphenyls (PCBs) were once commonly used electrical insulators and heat transfer agents. Their use has generally been phased out due to health concerns. PCBs were replaced by polybrominated diphenyl ethers (PBDEs), which bring similar toxicity and bioaccumulation concerns.

Toxicity

Some types of organochlorides have significant toxicity to plants or animals, including humans. Dioxins, produced when organic matter is burned in the presence of chlorine, and some insecticides, such as DDT, are persistent organic pollutants which pose dangers when they are released into the environment. For example, DDT, which was widely used to control insects in the mid 20th century, also accumulates in food chains, and causes reproductive problems (e.g., eggshell thinning) in certain bird species. Some organochlorine compounds, such as sulfur mustards, nitrogen mustards, and Lewisite, are even used as chemical weapons due to their toxicity.

However, the presence of chlorine in an organic compound does not ensure toxicity. Some organochlorides are considered safe enough for consumption in foods and medicines. For example, peas and broad beans contain the natural chlorinated plant hormone 4-chloroindole-3-acetic acid (4-Cl-IAA); and the sweetener sucralose (Splenda) is widely used in diet products. As of 2004, at least 165 organochlorides had been approved worldwide for use as pharmaceutical drugs, including the natural antibiotic vancomycin, the antihistamine loratadine (Claritin), the antidepressant sertraline (Zoloft), the anti-epileptic lamotrigine (Lamictal), and the inhalation anesthetic isoflurane.

Rachel Carson brought the issue of DDT pesticide toxicity to public awareness with her 1962 book *Silent Spring*. While many countries have phased out the use of some types of organochlorides such as the US ban on DDT, persistent DDT, PCBs, and other organochloride residues continue to be found in humans and mammals across the planet many years after production and use have been limited. In Arctic areas, particularly high levels are found in marine mammals. These chemicals concentrate in mammals, and are even found in human breast milk. In some species of marine mammals, particularly those that produce milk with a high fat content, males typically have far higher levels, as females reduce their concentration by transfer to their offspring through lactation.

Dioxins and Dioxin-like Compounds

Dioxins and dioxin-like compounds (DLCs) are compounds that are highly toxic environmental persistent organic pollutants (POPs). They are mostly by-products of various industrial processes - or, in case of dioxin-like PCBs and PBBs, part of intentionally produced mixtures. They include:

- Polychlorinated dibenzo-*p*-dioxins (PCDDs), or simply dioxins. PCDDs are derivatives of dibenzo-*p*-dioxin. There are 75 PCDD congeners (isomers), differing in the number and location of chlorine atoms, and seven of them are especially toxic.

- Polychlorinated dibenzofurans (PCDFs), or furans. PCDFs are derivatives of dibenzofuran. There are 135 isomers, ten have dioxin-like properties.

- Polychlorinated/polybrominated biphenyls (PCBs/PBBs), derived from biphenyl, of which twelve are "dioxin-like". Under certain conditions PCBs may form dibenzofurans/dioxins through partial oxidation.

- Finally, dioxin may refer to 1,4-Dioxin proper, the basic chemical unit of the more complex dioxins. This simple compound is not persistent and has no PCDD-like toxicity.

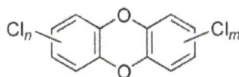

Polychlorinated dibenzo-*p*-dioxins

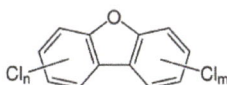

Polychlorinated dibenzofurans

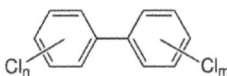

Polychlorinated biphenyls

1,4-dioxin

Because dioxins refer to such a broad class of compounds that vary widely in toxicity, the concept of toxic equivalence (TEQ) has been developed to facilitate risk assessment and regulatory control. Toxic equivalence factors (TEFs) exist for seven congeners of dioxins, ten furans and twelve PCBs. The reference congener is the most toxic dioxin 2,3,7,8-tetrachlorodibenzo-*p*-dioxin (TCDD) which per definition has a TEF of one.

In reference to their importance as environmental toxicants the term dioxins is used almost exclusively to refer to the sum of compounds (as TEQ) from the above groups which demonstrate the same specific toxic mode of action associated with TCDD. These include 17 PCDD/Fs and 12 PCBs. Incidents of contamination with PCBs are also often reported as dioxin contamination incidents since it is this toxic characteristic which is of most public and regulatory concern.

Toxicity

Mechanism of Toxicity

The toxic effects of dioxins are measured in fractional equivalencies of TCDD (2,3,7,8-tetrachlorodibenzo-*p*-dioxin), the most toxic and best studied member of its class. The toxicity is mediated through the interaction with a specific intracellular protein, the aryl hydrocarbon (AH) receptor, a transcriptional enhancer, affecting a number of other regulatory pro-teins. This receptor is a transcription factor which is involved in expression of many genes. TCDD binding to the AH receptor induces the cytochrome P450 1A class of enzymes which function to break down toxic compounds, e.g., carcinogenic polycyclic hydrocarbons such as benzo(a)pyrene (but making many of them more toxic in the process).

While the affinity of dioxins and related industrial toxicants to this receptor may not fully explain all their toxic effects including immunotoxicity, endocrine effects and tumor promotion, toxic responses appear to be typically dose-dependent within certain concentration ranges. A multiphasic dose-response relationship has also been reported, leading to uncertainty and debate about the true role of dioxins in cancer rates.

The endocrine disrupting activity of dioxins is thought to occur as a down-stream function of AH receptor activation, with thyroid status in particular being a sensitive marker of exposure. It is important to note that TCDD, along with the other PCDDs, PCDFs and dioxin-like coplanar PCBs are not direct agonists or antagonists of hormones, and are not active in assays which directly screen for these activities such as ER-CALUX and AR-CALUX. These compounds have also not been shown to have any direct mutagenic or genotoxic activity. Their main action in causing cancer is cancer promotion. A mixture of PCBs such as Aroclor may contain PCB compounds which are known estrogen agonists, but on the other hand are not classified as dioxin-like in terms of toxicity. Mutagenic effects have been established for some lower chlorinated chemicals such as 3-chlorodibenzofuran, which is neither persistent nor an AH receptor agonist.

Toxicity in Animals

The symptoms reported to be associated with dioxin toxicity in animal studies are incredibly wide ranging, both in the scope of the biological systems affected and in the range of dosage needed to bring these about. Acute effects of single high dose dioxin exposure include wasting syndrome, and typically a delayed death of the animal in 1 to 6 weeks. By far most toxicity studies have been performed using 2,3,7,8-tetrachlorodibenzo-p-dioxin.

The LD_{50} of TCDD varies wildly between species and even strains of the same species, with the most notable disparity being between the seemingly similar species of hamster and guinea pig. The oral LD_{50} for guinea pigs is as low as 0.5 to 2 µg/kg body weight, whereas the oral LD_{50} for hamsters can be as high as 1 to 5 mg/kg body weight. Even between different mouse or rat strains there may be tenfold to thousandfold differences in acute toxicity. Many pathological findings are seen in the liver, thymus and other organs.

Some chronic and sub-chronic exposures can be harmful at much lower levels, especially at particular developmental stages including foetal, neonatal and pubescent stages. Well established developmental effects are cleft palate, hydronephrosis, disturbances in tooth development and sexual development as well as endocrine effects.

Human Toxicity

Dioxins have been considered highly toxic and able to cause reproductive and developmental problems, damage the immune system, interfere with hormones and also cause cancer. This is based on animal studies. The best proven is chloracne. Even in poisonings with huge doses of TCDD, the only persistent effects after the initial malaise have been chloracne and amenorrhea. In occupational settings many symptoms have been seen, but exposures have always been to a multitude of chemicals including chlorophenols, chlorophenoxy acid herbicides, and solvents. Therefore, proof of dioxins as causative factors has been difficult. The suspected effects in adults are liver damage, and alterations in heme metabolism, serum lipid levels, thyroid functions, as well as diabetes and immunological effects.

In line with animal studies, developmental effects may be much more important than effects in adults. These include disturbances of tooth development, and of sexual development. An example of the variation in responses is clearly seen in a study following the Seveso disaster indicating that sperm count and motility were affected in different ways in exposed males, depending on whether they were exposed before, during or after puberty.

Intrauterine exposure to dioxins and dioxin-like compounds as an environmental toxin in pregnancy has subtle effects on the child later in life that include changes in liver function, thyroid hormone levels, white blood cell levels, and decreased performance in tests of learning and intelligence.

Exposure to dioxins can happen in a number of ways, most often as by-products of industrial waste. However, dioxins can result from natural processes including volcanic eruptions and forest fires, and manufacturing processes such as smelting, chlorine bleaching of paper pulp, and the creation of some herbicides and pesticides. Even at levels 100X lower than those associated with its cancer causing effects, the presence of dioxin can cause immune system damage, severe reproductive and developmental problems, and interference with regulatory hormones.

The Endometriosis Research Center (ERC) has testified before the California State Legislature concerning Assembly Bill 2820 [Cardoza, D-Merced] that, "feminine hygiene products (i.e. tampons) do indeed test positive for Dioxin. Dioxin, in turn, is a well-documented catalyst for Endometriosis - and the effects of Dioxin are cumulative; able to be measured as much as 20 or 30 years after exposure." The ERC also references an independent study that found, in an assessment of four brands of tampons and four brands of baby diapers, dioxins "were present at detectable concentrations in all samples." The presence of this toxin in tampons may be linked to endometriosis because dioxins last a long time in the body; they are chemically stable and can be absorbed by fat tissue, where they are then stored in the body. Their half-life in the body is estimated to be 7 to 11 years.

Carcinogenicity

Dioxins are well established carcinogens in animal studies, although the precise mechanistic role is not clear. Dioxins are not mutagenic or genotoxic. The United States Environmental Protection Agency has categorised dioxin, and the mixture of substances associated with sources of dioxin toxicity as a "likely human carcinogen". The International Agency for Research on Cancer has classified TCDD as a human carcinogen (class 1) on the basis of clear animal carcinogenicity and limited human data, but was not able to classify other dioxins. It is thought that the presence of dioxin can accelerate the formation of tumours and adversely affect the normal mechanisms for inhibiting tumour growth, without actually instigating the carcinogenic event.

As with all toxic endpoints of dioxin, a clear dose-response relationship is very difficult to establish. After accidental or high occupational exposures there is evidence on human carcinogenicity. There is much controversy especially on cancer risk at low population levels of dioxins. Among fishermen with high dioxin concentrations in their bodies, cancer deaths were decreased rather than increased. Some researchers have also proposed that dioxin induces cancer progression through a very different mitochondrial pathway.

Risk Assessment

The uncertainty and variability in the dose-response relationship of dioxins in terms of their toxicity, as well as the ability of dioxins to bioaccumulate mean that the tolerable daily intake (TDI) of dioxin has been set very low, 1-4 pg/kg body weight per day, i.e. 7×10^{-11} to 2.8×10^{-10} g per 70-kg person per day, to allow for this uncertainty and ensure public safety in all instances. Specifically, the TDI has been assessed based on the safety of children born to mothers exposed all their lifetime prior to pregnancy to such a daily intake of dioxins. It is likely that the TDI for other population groups could be somewhat higher. The most important cause for differences in different assessments is carcinogenicity. If the dose-response of TCDD in causing cancer is linear, it might be a true risk. If the dose-response is of a threshold-type or J-shape, there is little or no risk at the present concentrations. Understanding the mechanisms of toxicity better is hoped to increase the reliability of risk assessment.

Controversy

Greenpeace and some other environmental groups have called for the chlorine industry to be phased out. However, chlorine industry supporters say that "banning chlorine would mean that millions of people in the third world would die from want of disinfected water". (Although critics point out the existence of alternative water disinfection methods).

Sharon Beder and others have argued that the dioxin controversy has been very political and that large companies have tried to play down the seriousness of the problems of dioxin. The companies involved have often said that the campaign against dioxin is based on "fear and emotion" and not on science.

In 2008, Chile experienced a pork crisis caused by high dioxin concentrations in their pork exports. The contamination was found to be due to zinc oxide used in pork feed, and caused reputational and financial losses for the country, as well as leading to the introduction of new food safety regulations.

Human Intake and Levels

Most intake of dioxin-like chemicals is from food of animal origin: meat, dairy products, or fish predominate, depending on the country. The daily intake of dioxins and dioxin-like PCBs as TEQ is of the order of 100 pg/day, i.e. 1-2 pg/kg/day. In many countries both the absolute and relative significance of dairy products and meat have decreased due to strict emission controls, and brought about the decrease of total intake. E.g. in the United Kingdom the total intake of PCDD/F in 1982 was 239 pg/day and in 2001 only 21 pg/day (WHO-TEQ). Since the half-lives are very long (for e.g. TCDD 7–8 years), the body burden will increase almost over the whole lifetime. Therefore, the concentrations may increase five- to tenfold from age 20 to age 60. For the same reason, short term higher intake such as after food contamination incidents, is not crucial unless it is extremely high or lasts for several months or years.

The highest body burdens were found in Western Europe in the 1970s and early 1980s, and the trends have been similar in the U.S. The most useful measure of time trends is concentration in breast milk measured over decades. In many countries the concentrations have decreased to about one tenth of those in the 1970s, and the total TEQ concentrations are now of the order of 10-30 pg/g fat (please note the units, pg/g is the same as ng/kg, or the non-standard expression ppt used sometimes in America). The decrease is due to strict emission controls and also to the control of concentrations in food. In the U.S. young adult female population (age group 20-39), the concentration was 9.7 pg/g lipid in 2001-2002 (geometric mean).

Certain professions such as subsistence fishermen in some areas are exposed to exceptionally high amounts of dioxins and related substances. This along with high industrial exposures may be the most valuable source of information on the health risks of dioxins.

Uses

Dioxins have no common uses. They are manufactured on a small scale for chemical and toxicological research, but mostly exist as by-products of industrial processes such as bleaching paper pulp, pesticide manufacture, and combustion processes such as waste incineration. The defoliant Agent Orange contained dioxins. The production and use of dioxins was banned by the Stockholm Convention in 2001.

Sources

Environmental Sources

PCB-compounds, always containing low concentrations of dioxin-like PCBs and PCDFs, were synthesized for various technical purposes. They have entered the environment through accidents such as fires or leaks from transformers or heat exchangers, or from PCB-containing products in landfills or during incineration. Because PCBs are somewhat volatile, they have also been transported long distances by air leading to global distribution including the Arctic.

PCDD/F-compounds were never synthesized for any purpose, except for small quantities for scientific research. Small amounts of PCDD/Fs are formed whenever organics, oxygen and chlorine are available at suitable temperatures. This is augmented by metal catalysts such as copper. The optimal temperature range is 400 °C to 700 °C. This means that formation is highest when organic material is burned in less-than-optimal conditions such as open fires, building fires, domestic fireplaces, and poorly operated and/or designed solid waste incinerators. Historically, municipal and medical waste incineration was the most important source of PCDD/Fs.

Other sources of PCDD/F include:

- Metal smelting and refining

- Chlorine bleaching of pulp and paper - historically important source of PCDD/Fs to waterways.

- Synthesis side products of several chemicals, especially PCBs, chlorophenols, chlorophenoxy acid herbicides and hexachlorophene.

- Uncontrolled combustion, particularly open burning of waste ("backyard barrel burning"), accidental fires, wildfires.

- (Historical) Engines using leaded fuel, which contained the additives 1,2-Dichloroethane and 1,2-Dibromoethane.

In Waste Incineration

Improvements and changes have been made to nearly all industrial sources to reduce PCDD/F production. In waste incineration, large amounts of publicity and concern surrounded dioxin-like compounds during the 1980s-1990s and continues to pervade the public consciousness, especially when new incineration and waste-to-energy facilities are proposed. As a result of these concerns, incineration processes have been improved with increased combustion temperatures (over 1000 °C), better furnace control, and sufficient residence time allotted to ensure complete oxidation of organic compounds. Ideally, an incineration process oxidizes all carbon to CO_2 and converts all chlorine to

HCl or inorganic chlorides prior to the gases passing through the temperature window of 700-400 °C where PCDD/F formation is possible. These substances cannot easily form organic compounds, and HCl is easily and safely neutralized in the scrubber while CO_2 is vented to the atmosphere. Inorganic chlorides are incorporated into the ash.

Scrubber and particulate removal systems manage to capture most of the PCDD/F which forms even in sophisticated incineration plants. These PCDD/Fs are generally not destroyed but moved into the fly ash. Catalytic systems have been designed which destroy vapor-phase PCDD/Fs at relatively low temperatures. This technology is often combined with the baghouse or SCR system at the tail end of an incineration plant.

European Union limits for concentration of dioxin-like compounds in the discharged flue gas is 0.1 ng/Nm³ TEQ.

Both in Europe and in U.S.A., the emissions have decreased dramatically since the 1980s, by even 90%. This has also led to decreases in human body burdens, which is neatly demonstrated by the decrease of dioxin concentrations in breast milk.

Open burning of waste (backyard barrel burning) has not decreased effectively, and in the U.S. it is now the most important source of dioxins. Total U.S. annual emissions decreased from 14 kg in 1987 to 1.4 kg in 2000. However, backyard barrel burning decreased only modestly from 0.6 kg to 0.5 kg, resulting in over one third of all dioxins in the year 2000 from backyard burning alone.

Low concentrations of dioxins have been found in some soils without any anthropogenic contamination. A puzzling case of milk contamination was detected in Germany. The source was found to be kaolin added to animal feed. Dioxins have been repeatedly detected in clays from Europe and USA since 1996, with contamination of clay assumed to be the result of ancient forest fires or similar natural events with concentration of the PCDD/F during clay sedimentation.

Environmental Persistence and Bioaccumulation

All groups of dioxin-like compounds are persistent in the environment. Neither soil microbes nor animals are able to break down effectively the PCDD/Fs with lateral chlorines (positions 2,3,7, and 8). This causes very slow elimination. Ultraviolet light is able to slowly break down these compounds. Lipophilicity (tendency to seek for fat-like environments) and very poor water solubility make these compounds move from water environment to living organisms having lipid cell structures. This is called bioaccumulation. Increase in chlorination increases both stability and lipophilicity. The compounds with the very highest chlorine numbers (e.g. octachlorodibenzo-p-dioxin) are, however, so poorly soluble that this hinders their bioaccumulation. Bioaccumulation is followed by biomagnification. Lipid-soluble compounds are first accumulated to microscopic organisms such as phytoplankton (plankton of plant character, e.g. algae). Phytoplankton is consumed by animal plankton, this by invertebrates such as insects, these by small fish, and further by large fish and seals. At every stage or trophic level, the concentration is higher, because the persistent chemicals are not "burned off" when the higher organism uses the fat of the prey organism to produce energy.

Due to bioaccumulation and biomagnification, the species at the top of the trophic pyramid are most vulnerable to dioxin-like compounds. In Europe, the white-tailed eagle and some species of seals have approached extinction due to poisoning by persistent organic pollutants. Likewise, in America, the population of bald eagles declined because of POPs causing thinning of eggs and other reproductive problems. Usually, the failure has been attributed mostly to DDT, but dioxins are also a possible cause of reproductive effects. Both in America and in Europe, many waterfowl have high concentrations of dioxins, but usually not high enough to disturb their reproductive success. Due to supplementary winter feeding and other measures also, the white-tailed eagle is recovering. Also, ringed seals in the Baltic Sea are recovering.

Humans are also at the top of the trophic pyramid, particularly newborns. Exclusively breastfed newborns were estimated to be exposed to a total of 800 pg TEQ/day, leading to an estimated body weight-based dose of 242 pg TEQ/kg/day. Due to a multitude of food sources of adult humans exposure is much less averaging at 1 pg TEQ/kg-day, and dioxin concentrations in adults are much less at 10-100 pg/g, compared with 9000 to 340,000 pg/g (TEQ in lipid) in eagles or seals feeding almost exclusively on fish.

Because of different physicochemical properties, not all congeners of dioxin-like compounds find their routes to human beings equally well. Measured as TEQs, the dominant congeners in human tissues are 2,3,7,8-TCDD, 1,2,3,7,8-PeCDD, 1,2,3,6,7,8-HxCDD and 2,3,4,7,8-PeCDF. This is very different from most sources where hepta- and octa-congeners may predominate. The WHO panel re-evaluating the TEF values in 2005 expressed their concern that emissions should not be uncritically measured as TEQs, because all congeners are not equally important. They stated that "when a human risk assessment is to be done from abiotic matrices, factors such as fate, transport, and bioavailability from each matrix be specifically considered".

All POPs are poorly water-soluble, especially dioxins. Therefore, ground water contamination has not been a problem, even in cases of severe contamination due to the main chemicals such as chlorophenols. In surface waters, dioxins are bound to organic and inorganic particles.

Fate of Dioxins in Human Body

The same features causing persistence of dioxins in the environment, also cause very slow elimination in humans and animals. Because of low water solubility, kidneys are not able to secrete them in urine as such. They should be metabolised to more water-soluble metabolites, but also metabolism especially in humans is extremely slow. This results in biological half-lives of several years for all dioxins. That of TCDD is estimated to be 7 to 8 years, and for other PCDD/Fs from 1.4 to 13 years, PCDFs on average slightly shorter than PCDDs.

Dioxins are absorbed well from the digestive tract, if they are dissolved in fats or oils (e.g. in fish or meat). On the other hand, dioxins tend to adsorb tightly to soil particles, and absorption may be quite low: 13.8% of the given dose of TEQs in contaminated soil was absorbed.

In mammalian organisms, dioxins are found mostly in fat. Concentrations in fat seem to be relatively similar, be it serum fat, adipose tissue fat, or milk fat. This permits measuring dioxin burden by analysing breast milk. Initially, however, at least in laboratory animals, after a single dose, high concentrations are found in the liver, but in a few days, adipose tissue will predominate. In rat liver, however, high doses cause induction of CYP1A2 enzyme, and this binds dioxins. Thus, depending on the dose, the ratio of fat and liver tissue concentrations may vary considerably in rodents.

Sources of Human Exposure

The most important source of human exposure is fatty food of animal origin, and breast milk. There is much variation between different countries as to the most important items. In U.S. and Central Europe, milk, dairy products and meat have been by far the most important sources. In some countries, notably in Finland and to some extent in Sweden, fish is important due to contaminated Baltic fish and very low intake from any other sources. In most countries, a significant decrease of dioxin intake has occurred due to stricter controls during the last 20 years.

Historically occupational exposure to dioxins has been a major problem. Dioxins are formed as important toxic side products in the production of PCBs, chlorophenols, chlorophenoxy acid herbicides, and other chlorinated organic chemicals. This caused very high exposures to workers in poorly controlled hygienic conditions. Many workers had chloracne. In a NIOSH study in the U.S., the average concentration of TCDD in exposed persons was 233 ng/kg (in serum lipid) while it was 7 ng/kg in unexposed workers, even though the exposure had been 15–37 years earlier. This indicates a huge previous exposure. In fact the exact back-calculation is debated, and the concentrations may have been even several times higher than originally estimated.

Handling and spraying of chlorophenoxy acid herbicides may also cause quite high exposures, as clearly demonstrated by the users of Agent Orange in the Malayan Emergency and in the Vietnam War. The highest concentrations were detected in nonflying enlisted personnel (e.g. filling the tanks of planes), although the variation was huge, 0 to 618 ng/kg TCDD (mean 23.6 ng/kg). Other occupational exposures (working at paper and pulp mills, steel mills and incinerators) have been remarkably lower.

Accidental exposures have been huge in some cases. The highest concentrations in people after the Seveso accident were 56,000 ng/kg, and the highest exposure ever recorded was found in Austria in 1998, 144,000 ng/kg. This is equivalent to a dose of 20 to 30 µg/kg TCDD, a dose that would be lethal to guinea pigs and some rat strains.

Exposure from contaminated soil is possible when dioxins are blown up in dust, or children eat dirt. Inhalation was clearly demonstrated in Missouri in the 1970s, when waste oils were used as dust suppressant in horse arenas. Many horses and other animals were killed due to poisoning. Dioxins are neither volatile nor water-soluble, and therefore exposure of human beings depends on direct eating of soil or production of dust which carries the chemical. Contamination of ground water or breathing vapour of the chemical are not likely to cause a significant exposure. Currently, in the US, there are 126 Superfund sites with a completed exposure pathway contaminated with dioxins.

Further, PCBs are known to pass through treatment plants and accumulate in sludge which is used on farm fields in certain countries. In 2011 in South Carolina, SCDHEC enacted emergency sludge regulations after PCBs were found to have been discharged to a waste treatment plant.

PCBs are also known to flush from industry and land (aka sludge fields) to contaminate fish, as they have up and down the Catawba River in North and South Carolina. State authorities have posted fish consumption advisories due to accumulation of PCBs in fish tissue.

TEF Values

All dioxin-like compounds share a common mechanism of action via the aryl hydrocarbon receptor (AHR), but their potencies are very different. This means that similar effects are caused by all of them, but much larger doses of some of them are needed than of TCDD. Binding to the AHR as well as persistence in the environment and in the organism depends on the presence of so-called "lateral chlorines", in case of dioxins and furans, chlorine substitutes in positions 2,3,7, and 8. Each additional non-lateral chlorine decreases the potency, but qualitatively the effects remain similar. Therefore, a simple sum of different dioxin congeners is not a meaningful measure of toxicity. To compare the toxicities of various congeners and to render it possible to make a toxicologically meaningful sum of a mixture, a toxicity equivalency (TEQ) concept was created.

Each congener has been given a toxicity equivalence factor (TEF). This indicates its relative toxicity as compared with TCDD. Most TEFs have been extracted from *in vivo* toxicity data on animals, but if these are missing (e.g. in case of some PCBs), less reliable *in vitro* data have been used. After multiplying the actual amount or concentration of a congener by its TEF, the product is the virtual amount or concentration of TCDD having effects of the same magnitude as the compound in question. This multiplication is done for all compounds in a mixture, and these "equivalents of TCDD" can then simply be added, resulting in TEQ, the amount or concentration of TCDD toxicologically equivalent to the mixture.

The TEQ conversion makes it possible to use all studies on the best studied TCDD to assess the toxicity of a mixture. This resembles the common measure of all alcoholic drinks: beer, wine and whiskey can be added together as absolute alcohol, and this sum gives the toxicologically meaningful measure of the total impact.

The TEQ only applies to dioxin-like effects mediated by the AHR. Some toxic effects (especially of PCBs) may be independent of the AHR, and those are not taken into account by using TEQs.

TEFs are also approximations with certain amount of scientific judgement rather than scientific facts. Therefore, they may be re-evaluated from time to time. There have been several TEF versions since the 1980s. The most recent re-assessment was by an expert group of the World Health organization in 2005.

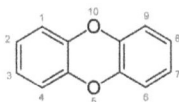

The skeletal formula and substituent numbering scheme of the parent compound dibenzo-*p*-dioxin

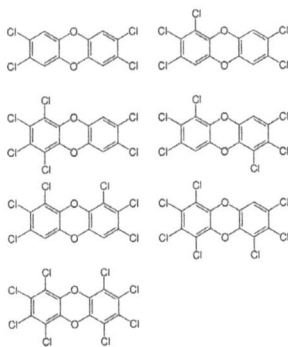

The 2,3,7,8-substituted PCDDs

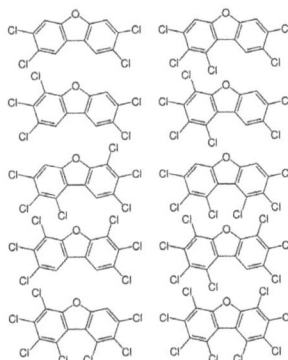

The 2,3,7,8-substituted PCDFs

Dioxin-like PCBs

WHO Toxic Equivalence Factors (WHO-TEF) for the dioxin-like congeners of concern

Polychlorinated dioxins	
2,3,7,8-TCDD	1
1,2,3,7,8-PeCDD	1
1,2,3,4,7,8-HxCDD	0.1
1,2,3,6,7,8-HxCDD	0.1
1,2,3,7,8,9-HxCDD	0.1
1,2,3,4,6,7,8-HpCDD	0.01
OCDD	0.0003
Polychlorinated dibenzofurans	
2,3,7,8-TCDF	0.1
1,2,3,7,8-PeCDF	0.03
2,3,4,7,8-PeCDF	0.3
1,2,3,4,7,8-HxCDF	0.1
1,2,3,6,7,8-HxCDF	0.1
1,2,3,7,8,9-HxCDF	0.1
2,3,4,6,7,8-HxCDF	0.1
1,2,3,4,6,7,8-HpCDF	0.01
1,2,3,4,7,8,9-HpCDF	0.01
OCDF	0.0003
Non-ortho-substituted PCBs	
3,3',4,4'-TCB (PCB77)	0.0001
3,4,4',5-TCB (PCB81)	0.0003

3,3',4,4',5-PeCB (PCB126)	0.1
3,3',4,4',5,5'-HxCB (PCB169)	0.03
Mono-ortho-substituted PCBs	
2,3,3',4,4'-PeCB (PCB105)	0.00003
2,3,4,4',5-PeCB (PCB114)	0.00003
2,3',4,4',5-PeCB (PCB118)	0.00003
2',3,4,4',5-PeCB (PCB123)	0.00003
2,3,3',4,4',5-HxCB (PCB156)	0.00003
2,3,3',4,4',5'-HxCB (PCB157)	0.00003
2,3',4,4',5,5'-HxCB (PCB167)	0.00003
2,3,3',4,4',5,5'-HpCB (PCB189)	0.00003

(T = tetra, Pe = penta, Hx = hexa, Hp = hepta, O = octa)

Dioxin Screening

There are two main methods for screening of dioxins and dioxin-like compounds:

- CALUX, or Chemical Activated Luciferase gene eXpression is a novel High-throughput screening bioassay. In comparison to HRGC-MS, it's a much faster and cheaper method as it is not reliant on expensive machinery used in HRGC-MS.

- HRGC-MS, or High Resolution Gas Chromatography Mass Spectrometry was the first screening method for 29 dioxin and DLC congeners.

Polychlorinated Dibenzodioxins

Polychlorinated dibenzodioxins (PCDDs), or simply dioxins, are a group of polyhalogenated organic compounds that are significant environmental pollutants.

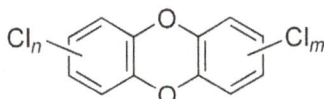

General structure of PCDDs where n and m can range from 0 to 4

They are commonly but inaccurately referred to as dioxins for simplicity, because every PCDD molecule contains a dibenzo-1,4-dioxin skeletal structure, with 1,4-dioxin as the central ring. Members of the PCDD family bioaccumulate in humans and wildlife because of their lipophilic properties, and may cause developmental disturbances and cancer.

Dioxins occur as by-products in the manufacture of some organochlorides, in the incineration of chlorine-containing substances such as polyvinyl chloride (PVC), in the chlorine bleaching of paper, and from natural sources such as volcanoes and forest fires. There have been many incidents of dioxin pollution resulting from industrial emissions and accidents; the earliest such incidents were in the mid 19th century during the Industrial Revolution.

The word "dioxins" may also refer to other similarly acting chlorinated compounds.

Chemical Structure of Dibenzo-1,4-dioxins

The structure of dibenzo-1,4-dioxin consists of two benzene rings joined by two oxygen bridges. This makes the compound an aromatic diether. The name dioxin formally refers to the central dioxygenated ring, which is stabilized by the two flanking benzene rings.

In PCDDs, chlorine atoms are attached to this structure at any of 8 different places on the molecule, at positions 1–4 and 6–9. There are 75 different PCDD congeners (that is, related dioxin compounds).

The toxicity of PCDDs depends on the number and positions of the chlorine atoms. Congeners that have chlorine in the 2, 3, 7, and 8 positions have been found to be significantly toxic. In fact, 7 congeners have chlorine atoms in the relevant positions which were considered toxic by the World Health Organization toxic equivalent (WHO-TEQ) scheme.

Historical Perspective

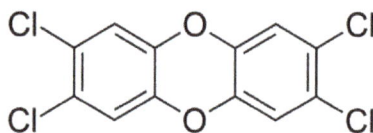

Structure of *2,3,7,8-tetrachlorodibenzodioxin* (TCDD)

Low concentrations of dioxins existed in nature prior to industrialization as a result of natural combustion and geological processes. Dioxins were first unintentionally produced as by-products from 1848 onwards as Leblanc process plants started operating in Germany. The first intentional synthesis of chlorinated dibenzodioxin was in 1872. Today, concentrations of dioxins are found in all humans, with higher levels commonly found in persons living in more industrialized countries. The most toxic dioxin, 2,3,7,8-tetrachlorodibenzodioxin (TCDD), became well known as a contaminant of Agent Orange, a herbicide used in the Malayan Emergency and the Vietnam War. Later, dioxins were found in Times Beach, Missouri and Love Canal, New York and Seveso, Italy. More recently, dioxins have been in the news with the poisoning of President Viktor Yushchenko of Ukraine in 2004, the Naples Mozzarella Crisis the 2008 Irish pork crisis, and the German feed incident of 2010.

Sources of Dioxins

The United States Environmental Protection Agency inventory of sources of dioxin-like compounds is possibly the most comprehensive review of the sources and releases of dioxins, but other countries now have substantial research as well.

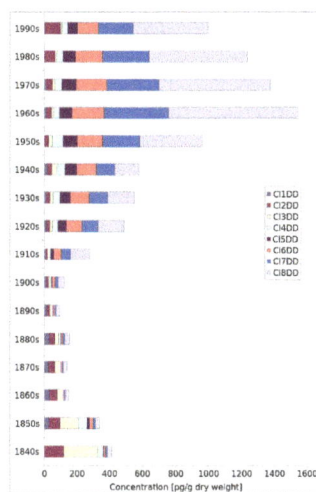

Concentration profile of PCDD in a dated sediment core from Esthwaite
Water, Cumbria, UK.

Occupational exposure is an issue for some in the chemical industries, historically for those making chlorophenols or chlorophenoxy acid herbicides or in the application of chemicals, notably herbicides. In many developed nations there are now emissions regulations which have dramatically decreased the emissions and thus alleviated some concerns, although the lack of continuous sampling of dioxin emissions causes concern about the understatement of emissions. In Belgium, through the introduction of a process called AMESA, continuous sampling showed that periodic sampling understated emissions by a factor of 30 to 50 times. Few facilities have continuous sampling.

Dioxins are produced in small concentrations when organic material is burned in the presence of chlorine, whether the chlorine is present as chloride ions or as organochlorine compounds, so they are widely produced in many contexts. According to the most recent US EPA data, the major sources of dioxins are broadly in the following types:

- Combustion sources, e.g. municipal waste or medical waste incinerators and private backyard barrel burning

- Metal smelting

- Refining and process sources

- Chemical manufacturing sources

- Natural sources

- Environmental reservoirs

When first carried out in 1987, the original US EPA inventory of dioxin sources revealed that incineration represented more than 80% of known dioxin sources. As a result, US EPA implemented new emissions requirements. These regulations succeeded in reducing dioxin stack emissions from incinerators. Incineration of municipal solid waste, medical waste, sewage sludge, and hazardous waste together now produce less than 3% of all dioxin emissions. Since 1987, however, backyard barrel burning has showed almost no decrease, and is now the largest source of dioxin emissions, producing about one third of the total output.

In incineration, dioxins can also reform or form *de novo* in the atmosphere above the stack as the exhaust gases cool through a temperature window of 600 to 200 °C. The most common method of reducing the quantity of dioxins reforming or forming *de novo* is through rapid (30 millisecond) quenching of the exhaust gases through that 400 °C window. Incinerator emissions of dioxins have been reduced by over 90% as a result of new emissions control requirements. Incineration in developed countries is now a very minor contributor to dioxin emissions.

Dioxins are also generated in reactions that do not involve burning — such as chlorine bleaching fibers for paper or textiles, and in the manufacture of chlorinated phenols, particularly when reaction temperature is not well controlled. Compounds involved include the wood preservative pentachlorophenol, and also herbicides such as 2,4-dichlorophenoxyacetic acid (or 2,4-D) and 2,4,5-trichlorophenoxyacetic acid (2,4,5-T). Higher levels of chlorination require higher reaction temperatures and greater dioxin production. Dioxins may also be formed during the photochemical breakdown of the common antimicrobial compound triclosan.

Sources of Human Intake

Tolerable daily, monthly or annual intakes have been set by the World Health Organization and a number of governments. Dioxins enter the general population almost exclusively from ingestion of food, specifically through the consumption of fish, meat, and dairy products since dioxins are fat-soluble and readily climb the food chain.

Children are passed substantial body burdens by their mothers, and breastfeeding increases the child's body burden. Dioxin exposure can also occur from contact with Pentachlorophenol (Penta) treated lumber as Pentachlorophenol often contains dioxins as a contaminant.Children's daily intakes during breast feeding are often many times above the intakes of adults based on body weight. This is why the WHO consultation group assessed the tolerable intake so as to prevent a woman from accumulating harmful body burdens before her first pregnancy. Breast fed children usually still have higher dioxin body burdens than non breast fed children. The WHO still recommends breast feeding for its other benefits. In many countries dioxins in breast milk have decreased by even 90% during the two last decades.

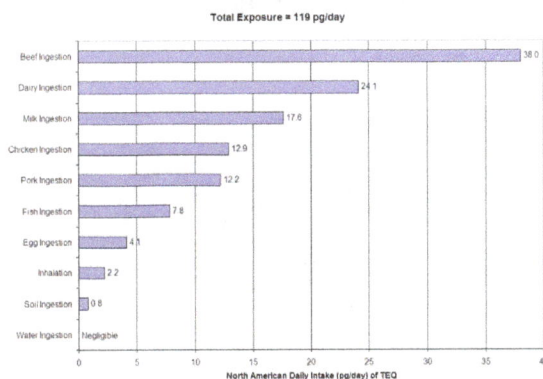

A chart illustrating how much dioxin the average American consumes per day. (Note: pg = picogram, or one trillionth of a gram, or 10^{-12} g) .

Dioxins are present in cigarette smoke. Dioxin in cigarette smoke was noted as "understudied" by the US EPA in its "Re-Evaluating Dioxin" (1995). In that same document, the US EPA acknowledged that dioxin in cigarettes is "anthropogenic" (man-made, "not likely in nature").

Metabolism

Dioxins are absorbed primarily through dietary intake of fat, as this is where they accumulate in animals and humans. In humans, the highly chlorinated dioxins are stored in fatty tissues and are neither readily metabolized nor excreted. The estimated elimination half-life for highly chlorinated dioxins (4–8 chlorine atoms) in humans ranges from 4.9 to 13.1 years.

The persistence of a particular dioxin congener in an animal is thought to be a consequence of its structure. Dioxins with no lateral (2, 3, 7, and 8) chlorines, which thus contain hydrogen atoms on adjacent pairs of carbons, can more readily be oxidized by cytochromes P450. The oxidized dioxins can then be more readily excreted rather than stored for a long time.

Toxicity

2,3,7,8-Tetrachlorodibenzodioxin (TCDD) is considered the most toxic of the congeners (for the mechanism of action, see 2,3,7,8-Tetrachlorodibenzodioxin and

Aryl hydrocarbon receptor). Other dioxin congeners including PCDFs and PCBs with dioxin-like toxicity, are given a toxicity rating from 0 to 1, where TCDD = 1. This toxicity rating is called the Toxic Equivalence Factor concept, or TEF. TEFs are consensus values and, because of the strong species dependence for toxicity, are listed separately for mammals, fish, and birds. TEFs for mammalian species are generally applicable to human risk calcula-tions. The TEFs have been developed from detailed assessment of literature data to facilitate both risk assessment and regulatory control. Many other compounds may also have dioxin-like properties, particularly non-ortho PCBs, one of which has a TEF as high as 0.1.

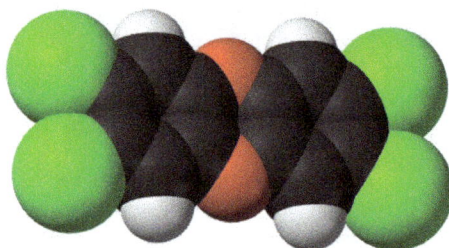

Space-filling model of 2,3,7,8-tetrachlorodibenzodioxin.

The total dioxin toxic equivalence (TEQ) value expresses the toxicity as if the mixture were pure TCDD. The TEQ approach and current TEFs have been adopted internation-ally as the most appropriate way to estimate the potential health risks of mixture of dioxins. Recent data suggest that this type of simple scaling factor may not be the most appropriate treatment for complex mixtures of dioxins; both transfer from the source and absorption and elimination vary among different congeners, and the TEF value is not able to accurately reflect this.

Dioxins and other persistent organic pollutants (POPs) are subject to the Stockholm Convention. The treaty obliges signatories to take measures to eliminate where possi-ble, and minimize where not possible to eliminate, all sources of dioxin.

Health Effects in Humans

Dioxins build up primarily in fatty tissues over time (bioaccumulation), so even small exposures may eventually reach dangerous levels. In 1994, the US EPA re-ported that dioxins are a probable carcinogen, but noted that non-cancer effects (reproduction and sexual development, immune system) may pose a greater threat to human health. TCDD, the most toxic of the dibenzodioxins, is classified as a Group 1 carcinogen by the International Agency for Research on Cancer (IARC). TCDD has a half-life of approximately 8 years in humans, although at high con-centrations, the elimination rate is enhanced by metabolism. The health effects of dioxins are mediated by their action on a cellular receptor, the aryl hydrocarbon receptor (AhR).

Exposure to high levels of dioxins in humans causes a severe form of persistent acne, known as chloracne. High occupational or accidental levels of exposures to dioxins have been shown by epidemiological studies to lead to an increased risk of tumors at all sites. Other effects in humans (at high dose levels) may include:

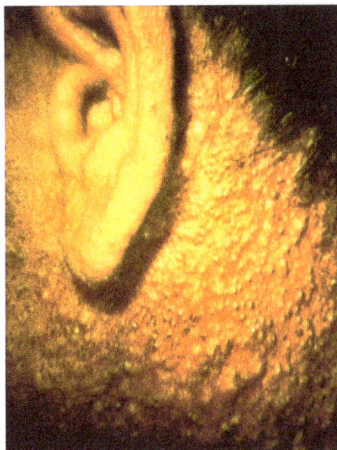

Chloracne on the ear and neck of a herbicide production worker.

- Developmental abnormalities in the enamel of children's teeth.

- Central and peripheral nervous system pathology

- Thyroid disorders

- Damage to the immune systems

- Endometriosis

- Diabetes

Recent studies have shown that high exposure to dioxins changes the ratio of male to female births among a population such that more females are born than males.

Dioxins accumulate in food chains in a fashion similar to other chlorinated compounds (bioaccumulation). This means that even small concentrations in contaminated water can be concentrated up a food chain to dangerous levels because of the long biological half life and low water solubility of dioxins.

Toxic Effects in Animals

While it has been difficult to establish specific health effects in humans due to the lack of controlled dose experiments, studies in animals have shown that dioxin causes a wide variety of toxic effects. In particular, TCDD has been shown to be teratogenic, mutagenic, carcinogenic, immunotoxic, and hepatotoxic. Furthermore, alterations in multiple endocrine and growth factor systems have been reported. The most sensitive effects, observed in multiple species, appear to be developmental, including effects on the developing immune, nervous, and reproductive systems. The most sensitive effects are caused at body burdens relatively close to those reported in humans.

Among the animals for which TCDD toxicity has been studied, there is strong evidence for the following effects:

- Birth defects (teratogenicity)

  In rodents, including rats, mice, hamsters and guinea pigs, birds, and fish.

- Cancer (including neoplasms in the mammalian lung, oral/nasal cavities, thyroid and adrenal glands, and liver, squamous cell carcinoma, and various animal hepatocarcinomas)

  In rodents and fish

- Hepatotoxicity (liver toxicity)

  In rodents, chickens, and fish

- Endocrine disruption

  In rodents and fish

- Immunosuppression

  In rodents and fish.

The LD_{50} of dioxin also varies wildly between species with the most notable disparity being between the ostensibly similar species of hamster and guinea pig. The oral LD_{50} for guinea pigs is as low as 0.5 to 2 µg/Kg body weight, whereas the oral LD_{50} for hamsters can be as high as 1 to 5 mg/Kg body weight, a difference of as much as thousandfold or more, and even among rat strains there may be thousandfold differences.

Agent Orange

Agent Orange was the code name for one of the herbicides and defoliants the U.S. military as part of its herbicidal warfare program, Operation Ranch Hand, during the Vietnam War from 1961 to 1971. It was a mixture of 2,4,5-T and 2,4-D. The 2,4,5-T used was contaminated with 2,3,7,8-tetrachlorodibenzodioxin (TCDD), an extremely toxic dioxin compound.

During the Vietnam war, between 1962 and 1971, the United States military sprayed 20,000,000 U.S. gallons (76,000,000 L) of chemical herbicides and defoliants in Vietnam, eastern Laos and parts of Cambodia, as part of Operation Ranch Hand.

By 1971, 12% of the total area of South Vietnam had been sprayed with defoliating chemicals, which were often applied at rates that were 13 times as high as the legal USDA limit. In South Vietnam alone, an estimated 10 million hectares of agricultural land were ultimately destroyed. In some areas, TCDD concentrations in soil and water were hundreds of times greater than the levels considered safe by the U.S. Environmental Protection Agency.

U.S. Army Huey helicopter spraying Agent Orange over Vietnamese agricultural land

According to Vietnamese Ministry of Foreign Affairs, 4.8 million Vietnamese people were exposed to Agent Orange, resulting in 400,000 people being killed or maimed, and 500,000 children born with birth defects. The Red Cross of Vietnam estimates that up to 1 million people are disabled or have health problems due to contaminated Agent Orange. The United States government has challenged these figures as being unreliable and unrealistically high.

Dioxin Exposure Incidents

- In 1949, in a Monsanto herbicide production plant for 2,4,5-T in Nitro, West Virginia, 240 people were affected when a relief valve opened.

- In 1963, a dioxin cloud escaped after an explosion in a Philips-Duphar plant (now Solvay Group) near Amsterdam. The plant was so polluted with dioxin after the accident that it had to be dismantled, embedded in concrete, and dumped into the ocean.

Spolana Neratovice chloralkali plant, air view

Between 1965 and 1968 production of 2,4,5-trichlorophenol in Spolana Neratovice plant in Czechoslovakia seriously poisoned about 60 workers with dioxins; after 3 years of investigations of the health problems of workers, Spolana
stopped manufacture of 2,4,5-T (most of which was supplied to the US military in Vietnam). Several buildings of the Spolana chemical plant were heavily contaminated by dioxins. Unknown amounts of dioxins were flushed into
the Elbe and Mulde rivers during the 2002 European flood, contaminating the
soils. Analysis of eggs and ducks found levels of dioxins 15-time higher than
EU limit and high concentrations of dioxin-like PCBs in the village of Libiš.
In 2004, the state health authority published a study which analysed the level of toxic substances in human blood near Spolana. According to the study,
blood dioxin levels in Neratovice, Libiš and Tišice were about twice the level of
the control group in Benešov. The quantity of dioxin chemicals near Spolana is
significantly higher than the background level in other countries, e.g., the US,
Japan or Spain. According to the US EPA, even the background level can pose a
risk of cancer from 1:10000 up to 1:1000, about 100 times higher than normal.
The consumption of local fish, eggs, poultry and some produce was prohibited
because of the post-flood contamination.

- Also during 1965 through 1968, Dr. Albert M. Kligman was contracted by the Dow
  Chemical Company to perform threshold tests for TCDD on inmates at Holmesburg Prison in Philadelphia after Dow studies revealed adverse effects on workers
  at Dow's Midland, Michigan plant were likely due to TCDD. A subsequent test by
  Dow in rabbit ear models when exposed to 4–8µg usually caused a severe response.
  The human studies carried out in Holmesburg failed to follow Dow's original protocol and lacked proper informed consent by the participants. As a result of poor
  study design and subsequent destruction of records, the tests were virtually worth-
  less even though ten inmates were exposed to 7,500µg of TCDD.

- In 1976, large amounts of dioxins were released in an industrial accident at
  Seveso, Italy, although no immediate human fatalities or birth defects occurred.

- In 1978, dioxins were some of the contaminants that forced the evacuation of
  the Love Canal neighborhood of Niagara Falls, New York.

- From 1982 through to 1985, Times Beach, Missouri, was bought out and evac-
  uated under order of the United States Environmental Protection Agency due
  to high levels of dioxins in the soil caused by applications of contaminated oil
  meant to control dust on the town's dirt roads. The town eventually disincor-
  porated.

- In December 1991, an electrical explosion caused dioxins (created from the oxidation of polychlorinated biphenyl) to spread through four residence halls and
  two other buildings on the college campus of SUNY New Paltz.

- In May 1999, there was a dioxin crisis in Belgium: quantities of polychlorinated biphenyls with dioxin-like toxicity had entered the food chain through contaminated animal feed. 7,000,000 chickens and 60,000 pigs had to be slaughtered. This scandal was followed by a landslide change in government in the elections one month later.

- Explosions resulting from the terrorist attacks on the US on September 11, 2001, released massive amounts of dust into the air. The air was measured for dioxins from September 23, 2001, to November 21, 2001, and reported to be "likely the highest ambient concentration that have ever been reported [in history]." The United States Environmental Protection Agency report dated October 2002 and released in December 2002 titled "Exposure and Human Health Evaluation of Airborne Pollution from the World Trade Center Disaster" authored by the EPA Office of Research and Development in Washington states that dioxin levels recorded at a monitoring station on Park Row near City Hall Park in New York between October 12 and 29, 2001, averaged 5.6 parts per trillion, or nearly six times the highest dioxin level ever recorded in the U.S. Dioxin levels in the rubble of the World Trade Centers were much higher with concentrations ranging from 10 to 170 parts per trillion. The report did no measuring of the toxicity of indoor air.

- In a 2001 case study, physicians reported clinical changes in a 30-year-old woman who had been exposed to a massive dosage (144,000 pg/g blood fat) of dioxin equal to 16,000 times the normal body level; the highest dose of dioxin ever recorded in a human. She suffered from chloracne, nausea, vomiting, epigastric pain, loss of appetite, leukocytosis, anemia, amenorrhoea and thrombocytopenia. However, other notable laboratory tests, such as immune function tests, were relatively normal. The same study also covered a second subject who had received a dosage equivalent to 2,900 times the normal level, who apparently suffered no notable negative effects other than chloracne. These patients were provided with olestra to accelerate dioxin elimination.

Viktor Yushchenko with chloracne after his TCDD poisoning incident

In 2004, in a notable individual case of dioxin poisoning, Ukrainian politician Viktor Yushchenko was exposed to the second-largest measured dose of dioxins, according to the reports of the physicians responsible for diagnosing him. This is the first known case of a single high dose of TCDD dioxin poisoning, and was diagnosed only after a toxicologist recognized the symptoms of chloracne while viewing television news coverage of his condition.

- In the early 2000s, residents of the city of New Plymouth, New Zealand, reported many illnesses of people living around and working at the Dow Chemical plant. This plant ceased production of 2,4,5-T in 1987.

- DuPont has been sued by 1,995 people who claim dioxin emissions from DuPont's plant in DeLisle, Mississippi, caused their cancers, illnesses or loved ones' deaths; of these only 850 were pending as of June 2008. In August 2005, Glen Strong, an oyster fisherman with the rare blood cancer multiple myeloma, was awarded $14 million from DuPont, but the ruling was overturned June 5, 2008, by a Mississippi jury who found DuPont's plant had no connection to Mr. Strong's disease. In another case, parents claimed dioxin from pollution caused the death of their 8-year-old daughter; the trial took place in the summer of 2007, and a jury wholly rejected the family's claims, as no scientific connection could be proven between DuPont and the family's tragic loss. DuPont's DeLisle plant is one of three titanium dioxide facilities (including Edgemoor, Delaware, and New Johnsonville, Tennessee) that are the largest producers of dioxin in the country, according to the US EPA's Toxic Release Inventory. DuPont maintains its operations are safe and environmentally responsible.

- In 2007, thousands of tonnes of foul-smelling refuse were piled up in Naples, Italy and its surrounding villages, defacing entire neighbourhoods. Authorities discovered that polychlorinated dibenzodioxins levels in buffalo milk used by 29 mozzarella makers exceeded permitted limits; after further investigation they impounded milk from 66 farms. Authorities suspected the source of the contamination was from waste illegally disposed of on land grazed by buffalo. Prosecutors in Naples placed 109 people under investigation on suspicion of fraud and food poisoning. Sales of Mozzarella cheese fell by 50% in Italy.

- In December 2008 in Ireland dioxin levels in pork were disclosed to have been between 80 and 200 times the legal limit. All Irish pork products were withdrawn from sale both nationally and internationally.

In this case the dioxin toxicity was found to be mostly due to dioxin-like polychlorinated dibenzofurans and polychlorinated biphenyls, and the contribution from actual polychlorinated dibenzodioxins was relatively low. It is thought that the incident resulted from the contamination of fuel oil used in a drying burner at a single feed processor, with PCBs. The resulting combustion produced a highly toxic mixture of PCBs, dioxins and furans, which was included in the feed produced and subsequently fed to a large number of pigs.

- According to the last available data, in 2005 the production of dioxin by the steel industry ILVA in Taranto (Italy) accounted for 90.3 per cent of the overall Italian emissions, and 8.8 per cent of the European emissions.

- German dioxin incident: In January 2011 about 4700 German farms were banned from making deliveries after self-checking of an animal feed producer had showed levels of dioxin above maximum levels. This incident appeared to involve PCDDs and not PCBs. Dioxins were found in animal feed and eggs in many farms. The maximum values were exceeded twofold in feed and maximally fourfold in some individual eggs. Thus the incident was minor as compared with the Belgian crisis in 1999, and delivery bans were rapidly cleared.

Dioxin Testing

The analyses used to determine these compounds' relative toxicity share common elements that differ from methods used for more traditional analytical determinations. The preferred methods for dioxins and related analyses use high resolution gas chromatography/mass spectrometry (HRGC/HRMS). Concentrations are determined by measuring the ratio of the analyte to the appropriate isotopically labeled internal standard.

Also novel bio-assays like DR CALUX are nowadays used in identification of dioxins and dioxin-like compounds. The advantage in respect to HRGC/HRMS is that it is able to scan many samples at lower costs. Also it is able to detect all compounds that interact with the Ah-receptor which is responsible for carcinogenic effects.

Polychlorinated Dibenzofurans

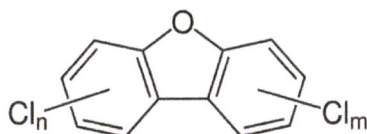

General chemical structure of PCDFs, where $2 \leq n+m \leq 8$

Polychlorinated dibenzofurans (PCDFs) are a family of organic compounds with one or several of the hydrogens in the dibenzofuran structure replaced by chlorines. For example, 2,3,7,8-tetrachlorodibenzofuran (TCDF) has chlorine atoms substituted for each of the hydrogens on the number 2, 3, 7, and 8 carbons. Polychlorinated dibenzofurans are much more toxic than the parent compounds, with properties and chemical structures similar to polychlorinated dibenzodioxins. These groups together are often inaccurately called dioxins. They are known teratogens, mutagens, and suspected human carcinogens. PCDFs tend to cooccur with polychlorinated dibenzodioxins (PCDDs). PCDFs can be formed by pyrolysis or incineration at temperatures below 1200 °C of chlorine containing products, such as PVC, PCBs, and other organochlorides, or of non-chlorine containing products in the presence of chlorine donors. Dibenzofurans are known persistent organic pollutants (POP), classified among the *dirty dozen* in the Stockholm Convention on Persistent Organic Pollutants.

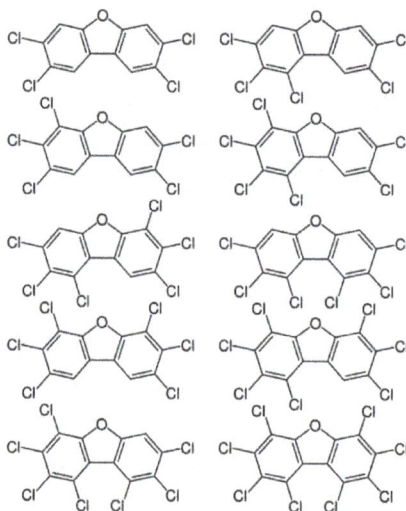

Structures of the ten 2,3,7,8-substituted PCDF congeners that are toxicologically of most relevance

Safety, Toxicity, Regulation

Occupational exposure to dibenzofuran may occur through inhalation and contact with the skin, particularly at sites where coal tar, coal tar derivatives, and creosote are produced or used.

Fish and dairy intake also have been studied in relation to the body burden of dibenzofuran in pregnant women. Consumption of contaminated water and food are the primary sources of exposure.

Dibenzofuran is excreted 22% the daily intake of dioxins from meals is excreted from feces and 29% from sebum.

Dibenzofuran is cited in the United States Clean Air Act 1990 Amendments -Hazardous Air Pollutants as a volatile hazardous air pollutant of potential concern. The Superfund Amendment Reauthorization Act (SARA) Section 110 placed dibenzofuran on the revised Agency for Toxic Substances and Disease Registry (ATSDR) priority list of hazardous substances to be the subject of a toxicological profile. The listing was based on the substance's frequency of occurrence at Comprehensive Environmental Response, Compensation, and Liability Act (CERCLA) National Priorities List sites, its toxicity, and/or its potential for human exposure. Dibenzofuran is also listed in the Massachusetts Substance List for Right-to-Know Law, the New Jersey Department of Health Hazard Right-to-Know Program Hazardous Substance List, and the Pennsylvania Department of Labor and Industry Hazardous Substance List. California's Air Toxics "Hot Spots" List (Assembly Bill).

1,4-Dioxin

1,4-Dioxin (also referred as dioxin or *p*-dioxin) is a heterocyclic, organic, non-aromatic compound with the chemical formula $C_4H_4O_2$. There is an isomeric form of 1,4-dioxin, 1,2-dioxin (or *o*-dioxin). 1,2-Dioxin is very unstable due to its peroxide-like characteristics.

The term "dioxin" is most commonly used for a family of derivatives of dioxin, known as polychlorinated dibenzodioxins (PCDDs).

Preparation

1,4-Dioxin can be prepared by cycloaddition, namely by the Diels–Alder reaction of furan and maleic anhydride. The adduct formed has a carbon-carbon double bond, which is converted to an epoxide. The epoxide then undergoes a retro-Diels–Alder reaction, forming 1,4-dioxin and regenerating maleic anhydride.

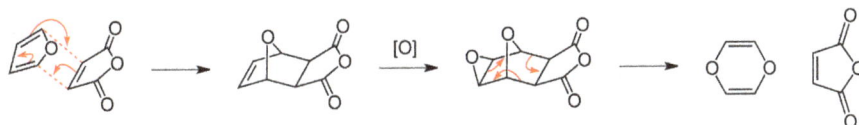

Derivatives

The word "dioxin" can refer in a general way to compounds which have a dioxin core skeletal structure with substituent molecular groups attached to it. For example, dibenzo-1,4-dioxin is a compound whose structure consists of two benzo- groups fused onto a 1,4-dioxin ring.

Polychlorinated Dibenzodioxins

Because of their extreme importance as environmental pollutants, current scientific literature uses the name **dioxins** commonly for simplification to denote the chlorinated derivatives of dibenzo-1,4-dioxin, more precisely the polychlorinated dibenzodioxins (PCDDs), among which 2,3,7,8-tetrachlorodibenzodioxin (TCDD), a tetrachlorinated derivative, is the best known. The polychlorinated dibenzodioxins, which can also be classified in the family of halogenated organic compounds, have been shown to bioaccumulate in humans and wildlife due to their lipophilic properties, and are known teratogens, mutagens, and carcinogens.

PCDDs are formed through combustion, chlorine bleaching and manufacturing processes. The combination of heat and chlorine creates dioxin. Since chlorine is often a part of the Earth's environment, natural ecological activity such as volcanic activity and forest fires can lead to the formation of PCDDs. Nevertheless, PCDDs are mostly produced by human activity.

Famous PCDD exposure cases include Agent Orange sprayed over vegetation by the British military in Malaya during the Malayan Emergency and the U.S. military in Vietnam during the Vietnam War, the Seveso disaster, and the poisoning of Viktor Yushchenko.

Polychlorinated dibenzofurans are a related class compounds to PCDDs which are often included within the general term "dioxins".

Polycyclic Aromatic Hydrocarbon

Polycyclic aromatic hydrocarbons (PAHs, also *polyaromatic hydrocarbons*) are hydrocarbons—organic compounds containing only carbon and hydrogen—that are composed of multiple aromatic rings (organic rings in which the electrons are delocalized). Formally, the class is further defined as lacking further branching substituents on these ring structures. Polynuclear aromatic hydrocarbons (PNAs) are a subset of PAHs that have fused aromatic rings, that is, rings that share one or more sides. The simplest such chemicals are naphthalene, having two aromatic rings, and the three-ring compounds anthracene and phenanthrene.

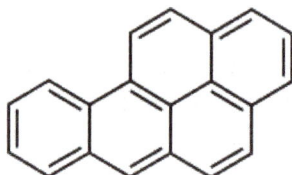

Standard line-angle schematic representation of an important PAH, benzo[*a*]pyrene, where carbon atoms are represented by the vertices of the hexagons, and hydrogens are inferred as projecting out at 120° angles to fill the fourth carbon valence (as necessary)

PAHs are neutral, nonpolar molecules found in coal and in tar deposits. They are produced as well by incomplete combustion of organic matter (e.g., in engines and incinerators, when biomass burns in forest fires, etc.).

PAHs may also be abundant in the universe, and are conjectured to have formed as early as the first couple of billion years after the Big Bang, in association with formation of new stars and exoplanets. Some studies suggest that PAHs account for a significant percentage of all carbon in the universe, and PAHs are discussed as possible starting materials for abiologic syntheses of materials required by the earliest forms of life.

Nomenclature, Structure, Properties

Nomenclature and Structure

The tricyclic species phenanthrene and anthracene represent the starting members of the PAHs. Smaller molecules, such as benzene, are not PAHs, and PAHs are not generally considered to contain heteroatoms or carry substituents.

PAHs with five or six-membered rings are most common. Those composed only of six-membered rings are called *alternant* PAHs, which include *benzenoid* PAHs. The following are examples of PAHs that vary in the number and arrangement of their rings:

- Examples of PAH compounds

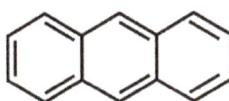
Anthracene

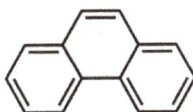
Phenanthrene

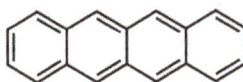
Tetracene

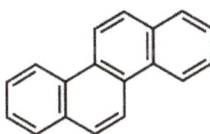
Chrysene

PAHs are nonpolar and lipophilic. PAHs are insoluble in water. The larger members are also poorly soluble in organic solvents as well as lipids. They are usually colorless.

Although PAHs clearly are aromatic compounds, the degree of aromaticity can be different for each ring segment. According to *Clar's rule* (formulated by Erich Clar in 1964) for PAHs the resonance structure with the most disjoint aromatic π-sextets—i.e. benzene-like moieties—is the most important for the characterization of the properties.

For example, in phenanthrene one Clar structure has two sextets at the extremities, while the other resonance structure has just one central sextet; therefore in this molecule the outer rings have greater aromatic character whereas the central ring is less aromatic and therefore more reactive. In contrast, in anthracene the resonance structures have one sextet, which can be at any of the three rings, and the aromaticity spreads out more evenly across the whole molecule. This difference in number of sextets is reflected in the UV absorbance spectra of these two isomers; phenanthrene has a highest wavelength absorbance around 290 nm, while anthracene has highest wavelength bands around 380 nm. Three Clar structures with two sextets each are present in chrysene. Superposition of these structures reveals that the aromaticity in the outer rings is greater (each has a sextet in two of the three Clar structures) compared to the inner rings (each has a sextet in only one of the three).

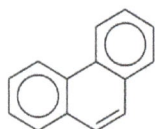 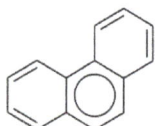   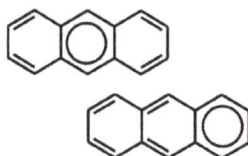   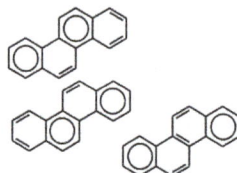

| Phenanthrene | Anthracene | Chrysene |

Sources & Distribution

Polycyclic aromatic hydrocarbons are primarily found in natural sources such as creosote. They can result from the incomplete combustion of organic matter. PAHs can also be produced geologically when organic sediments are chemically transformed into fossil fuels such as oil and coal. The dominant sources of PAHs in the environment are thus from human activity: Wood-burning and combustion of other biofuels such as dung or crop residues contribute more than half of annual global PAH emissions, particularly due to biofuel use in India and China. As of 2004, industrial processes and the extraction and use of fossil fuels made up slightly more than one quarter of global PAH emissions, dominating outputs in industrial countries such as the United States. Wild fires are another notable source. Substantially higher outdoor air, soil, and water concentrations of PAHs have been measured in Asia, Africa, and Latin America than in Europe, Australia, and the U.S./Canada.

PAHs are typically found as complex mixtures. Lower-temperature combustion, such as tobacco smoking or wood-burning, tends to generate low molecular weight PAHs, whereas high-temperature industrial processes typically generate PAHs with higher molecular weights.

In the Aqueous Environment

Most PAH's are insoluble in water, which limits their mobility in the environment. Aqueous solubility of PAHs decreases approximately logarithmically as molecular mass increases.

Two-ring PAHs, and to a lesser extent three-ring PAHs, dissolve in water, making them more available for biological uptake and degradation. Further, two- to four-ring PAHs volatilize sufficiently to appear in the atmosphere predominantly in gaseous form, although the physical state of four-ring PAHs can depend on temperature. In contrast, compounds with five or more rings have low solubility in water and low volatility; they are therefore predominantly in solid state, bound to particulate air pollution, soils, or sediments. In solid state, these compounds are less accessible for biological uptake or degradation, increasing their persistence in the environment.

Human Exposure

Human exposure varies across the globe and depends on factors such as smoking rates, fuel types in cooking, and pollution controls on power plants, industrial processes, and vehicles. Developed countries with stricter air and water pollution controls, cleaner sources of cooking (i.e., gas and electricity vs. coal or biofuels), and prohibitions of public smoking tend to have lower levels of PAH exposure, while developing and undeveloped countries tend to have higher levels.

A wood-burning open-air cook stove. Smoke from solid fuels like wood is a large source of PAHs globally.

Burning solid fuels such as coal and biofuels in the home for cooking and heating is a dominant global source of PAH emissions that in developing countries leads to high levels of exposure to indoor particulate air pollution containing PAHs, particularly for women and children who spend more time in the home or cooking.

In industrial countries, people who smoke tobacco products, or who are exposed to second-hand smoke, are among the most highly exposed groups; tobacco smoke contributes to 90% of indoor PAH levels in the homes of smokers. For the general population in developed countries, the diet is otherwise the dominant source of PAH exposure, particularly from smoking or grilling meat or consuming PAHs deposited on plant foods, especially broad-leafed vegetables, during growth. PAHs are typically at low concentrations in drinking water.

Emissions from vehicles such as cars and trucks can be a substantial outdoor source of PAHs in particulate air pollution. Geographically, major roadways are thus sources of PAHs, which may distribute in the atmosphere or deposit nearby. Catalytic converters are estimated to reduce PAH emissions from gasoline-fired vehicles by 25-fold.

Smog in Cairo, Egypt. Particulate air pollution, including smog, is a substantial avenue for human exposure to PAHs.

People can also be occupationally exposed during work that involves fossil fuels or their derivatives, wood burning, carbon electrodes, or exposure to diesel exhaust. Industrial activity that can produce and distribute PAHs includes aluminum, iron, and steel manufacturing; coal gasification, tar distillation, shale oil extraction; production of coke, creosote, carbon black, and calcium carbide; road paving and asphalt manufacturing; rubber tire production; manufacturing or use of metal working fluids; and activity of coal or natural gas power stations.

Environmental Distribution and Degradation

PAHs typically disperse from urban and suburban non-point sources through road run-off, sewage, and atmospheric circulation and subsequent deposition of particulate air pollution. Soil and river sediment near industrial sites such as creosote manufacturing facilities can be highly contaminated with PAHs. Oil spills, creosote, coal mining dust, and other fossil fuel sources can also distribute PAHs in the environment.

Oil on a beach after a 2007 oil spill in Korea.

Two- and three-ring PAHs can disperse widely while dissolved in water or as gases in the atmosphere, while PAHs with higher molecular weights can disperse locally or regionally adhered to particulate matter that is suspended in air or water until the particles land or settle out of the water column. PAHs have a strong affinity for organic carbon, and thus highly organic sediments in rivers, lakes, and the ocean can be a substantial sink for PAHs.

Algae and some invertebrates such as protozoans, mollusks, and many polychaetes have limited ability to metabolize PAHs and bioaccumulate disproportionate concentrations of PAHs in their tissues; however, PAH metabolism can vary substantially across invertebrate species. Most vertebrates metabolize and excrete PAHs relatively rapidly. Tissue concentrations of PAHs do not increase (biomagnify) from the lowest to highest levels of food chains.

PAHs transform slowly to a wide range of degradation products. Biological degradation by microbes is a dominant form of PAH transformation in the environment. Soil-consuming invertebrates such as earthworms speed PAH degradation, either through direct metabolism or by improving the conditions for microbial transformations. Abiotic degradation in the atmosphere and the top layers of surface waters can produce nitrogenated, halogenated, hydroxylated, and oxygenated PAHs; some of these compounds can be more toxic, water-soluble, and mobile than their parent PAHs.

Minor Sources

Volcanic eruptions may emit PAHs. Certain PAHs such as perylene can also be generated in anaerobic sediments from existing organic material, although it remains undetermined whether abiotic or microbial processes drive their production.

Human Health

Cancer is a primary human health risk of exposure to PAHs. Exposure to PAHs has also been linked with cardiovascular disease and poor fetal development.

Cancer

PAHs have been linked to skin, lung, bladder, liver, and stomach cancers in well-established animal model studies. Specific compounds classified by various agencies as possible or probable human carcinogens are identified in the section "Regulation and Oversight" below.

Historical Significance

Historically, PAHs contributed substantially to our understanding of adverse health effects from exposures to environmental contaminants, including chemical carcinogenesis. In 1775, Percivall Pott, a surgeon at St. Bartholomew's Hospital in London, observed that scrotal cancer was unusually common in chimney sweepers and proposed the cause as occupational exposure to soot. A century later, Richard von Volkmann reported increased skin cancers in workers of the coal tar industry of Germany, and by the early 1900s increased rates of cancer from exposure to soot and coal tar was widely accepted. In 1915, Yamigawa and Ichicawa were the first to experimentally produce cancers, specifically of the skin, by topically applying coal tar to rabbit ears.

In 1922, Ernest Kennaway determined that the carcinogenic component of coal tar mixtures was an organic compound consisting of only C and H. This component was later linked to a characteristic fluorescent pattern that was similar but not identical to benz[*a*]anthracene, a PAH that was subsequently demonstrated to cause tumors. Cook, Hewett and Hieger then linked the specific spectroscopic fluorescent profile of benzo[*a*]pyrene to that of the carcinogenic component of coal tar, the first time that a specific compound from an environmental mixture (coal tar) was demonstrated to be carcinogenic.

An eighteenth century drawing of chimney sweeps.

In the 1930s and later, epidemiologists from Japan, England, and the U.S., including Richard Doll and various others, reported greater rates of death from lung cancer following occupational exposure to PAH-rich environments among workers in coke ovens and coal carbonization and gasification processes.

Mechanisms of Carcinogenesis

The structure of a PAH influences whether and how the individual compound is carcinogenic. Some carcinogenic PAHs are genotoxic and induce mutations that initiate cancer; others are not genotoxic and instead affect cancer promotion or progression.

PAHs that affect cancer initiation are typically first chemically modified by enzymes into metabolites that react with DNA, leading to mutations. When the DNA sequence is altered in genes that regulate cell replication, cancer can result. Mutagenic PAHs, such as benzo[a]pyrene, usually have four or more aromatic rings as well as a "bay region", a structural pocket that increases reactivity of the molecule to the metabolizing enzymes. Mutagenic metabolites of PAHs include diol epoxides, quinones, and radical PAH cations. These metabolites can bind to DNA at specific sites, forming bulky complexes called DNA adducts that can be stable or unstable. Stable adducts may lead to DNA replication errors, while unstable adducts react with the DNA strand, removing a purine base (either adenine or guanine). Such mutations, if they are not repaired, can transform genes encoding for normal cell signaling proteins into cancer-causing oncogenes. Quinones can also repeatedly generate reactive oxygen species that may independently damage DNA.

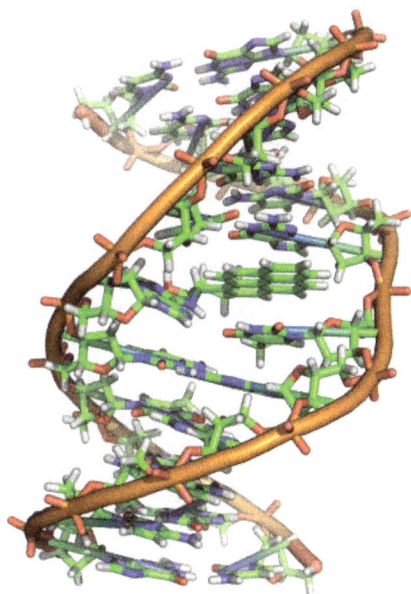

An adduct formed between a DNA strand and an epoxide derived from a benzo[a]pyrene molecule (center); such adducts may interfere with normal DNA replication.

Enzymes in the cytochrome family (CYP1A1, CYP1A2, CYP1B1) metabolize PAHs to diol epoxides. PAH exposure can increase production of the cytochrome enzymes, allowing the enzymes to convert PAHs into mutagenic diol epoxides at greater rates. In this pathway, PAH molecules bind to the aryl hydrocarbon receptor (AhR) and activate it as a transcription factor that increases production of the cytochrome enzymes. The activity of these enzymes may at times conversely protect against PAH toxicity, which is not yet well understood.

Low molecular weight PAHs, with 2 to 4 aromatic hydrocarbon rings, are more potent as co-carcinogens during the promotional stage of cancer. In this stage, an initiated cell (i.e., a cell that has retained a carcinogenic mutation in a key gene related to cell replication) is removed from growth-suppressing signals from its neighboring cells and begins to clonally replicate. Low molecular weight PAHs that have bay or bay-like regions can dysregulate gap junction channels, interfering with intercellular communication, and also affect mitogen-activated protein kinases that activate transcription factors involved in cell proliferation. Closure of gap junction protein channels is a normal precursor to cell division. Excessive closure of these channels after exposure to PAHs results in removing a cell from the normal growth-regulating signals imposed by its local community of cells, thus allowing initiated cancerous cells to replicate. These PAHs do not need to be enzymatically metabolized first. Low molecular weight PAHs are prevalent in the environment, thus posing a significant risk to human health at the promotional phases of cancer.

Cardiovascular Disease

Adult exposure to PAHs has been linked to cardiovascular disease. PAHs are among the complex suite of contaminants in cigarette smoke and particulate air pollution and may contribute to cardiovascular disease resulting from such exposures.

In laboratory experiments, animals exposed to certain PAHs have shown increased development of plaques (atherogenesis) within arteries. Potential mechanisms for the pathogenesis and development of atherosclerotic plaques may be similar to the mechanisms involved in the carcinogenic and mutagenic properties of PAHs. A leading hypothesis is that PAHs may activate the cytochrome enzyme CYP1B1 in vascular smooth muscle cells. This enzyme then metabolically processes the PAHs to quinone metabolites that bind to DNA in reactive adducts that remove purine bases. The resulting mutations may contribute to unregulated growth of vascular smooth muscle cells or to their migration to the inside of the artery, which are steps in plaque formation. These quinone metabolites also generate reactive oxygen species that may alter the activity of genes that affect plaque formation.

Oxidative stress following PAH exposure could also result in cardiovascular disease by causing inflammation, which has been recognized as an important factor in the development of atherosclerosis and cardiovascular disease. Biomarkers of exposure to PAHs in humans have been associated with inflammatory biomarkers that are recognized as important predictors of cardiovascular disease, suggesting that oxidative stress resulting from exposure to PAHs may be a mechanism of cardiovascular disease in humans.

Developmental Impacts

Multiple epidemiological studies of people living in Europe, the United States, and China have linked in utero exposure to PAHs, through air pollution or parental occupational exposure, with poor fetal growth, reduced immune function, and poorer neurological development, including lower IQ.

Regulation and Oversight

Some governmental bodies, including the European Union as well as NIOSH and the Environmental Protection Agency in the US, regulate concentrations of PAHs in air, water, and soil. The European Commission has restricted concentrations of 8 carcinogenic PAHs in consumer products that contact the skin or mouth.

Priority polycyclic aromatic hydrocarbons identified by the US EPA, the US Agency for Toxic Substances and Disease Registry (ATSDR), and the European Food Safety Authority (EFSA) due to their carcinogenicity or genotoxicity and/or ability to be monitored are the following:

Compound	Agency	Compound	Agency
acenaphthene	EPA, ATSDR	cyclopenta[c,d] pyrene	EFSA
acenaphthylene	EPA, ATSDR	dibenz[a,h]anthra-cene[A]	EPA, ATSDR, EFSA
anthracene	EPA, ATSDR	dibenzo[a,e]pyrene	EFSA
benz[a]anthracene[A]	EPA, ATSDR, EFSA	dibenzo[a,h]pyrene	EFSA
benzo[b]fluoran-thene[A]	EPA, ATSDR, EFSA	dibenzo[a,i]pyrene	EFSA
benzo[j]fluoranthene	ATSDR, EFSA	dibenzo[a,l]pyrene	EFSA
benzo[k]fluoran-thene[A]	EPA, ATSDR, EFSA	fluoranthene	EPA, ATSDR
benzo[c]fluorene	EFSA	fluorene	EPA, ATSDR
benzo[g,h,i] perylene[A]	EPA, ATSDR, EFSA	indeno[1,2,3-c,d] pyrene[A]	EPA, ATSDR, EFSA
benzo[a]pyrene[A]	EPA, ATSDR, EFSA	5-methylchrysene	EFSA
benzo[e]pyrene	ATSDR	naphthalene	EPA
chrysene[A]	EPA, ATSDR, EFSA	phenanthrene	EPA, ATSDR
coronene	ATSDR	pyrene	EPA, ATSDR

[A] Considered probable or possible human carcinogens by the US EPA, the European Union, and/or the International Agency for Research on Cancer (IARC).

Detection and Optical Properties

A spectral database exists for tracking polycyclic aromatic hydrocarbons (PAHs) in the universe.greatly upgraded database Detection of PAHs in materials is often done using gas chromatography-mass spectrometry or liquid chromatography with ultraviolet-visible or fluorescence spectroscopic methods or by using rapid test PAH indicator strips.

PAHs possess very characteristic UV absorbance spectra. These often possess many absorbance bands and are unique for each ring structure. Thus, for a set of isomers, each isomer has a different UV absorbance spectrum than the others. This is particularly useful in the identification of PAHs. Most PAHs are also fluorescent, emitting characteristic wavelengths of light when they are excited (when the molecules absorb light). The extended pi-electron electronic structures of PAHs lead to these spectra, as well as to certain large PAHs also exhibiting semi-conducting and other behaviors.

Origins of Life

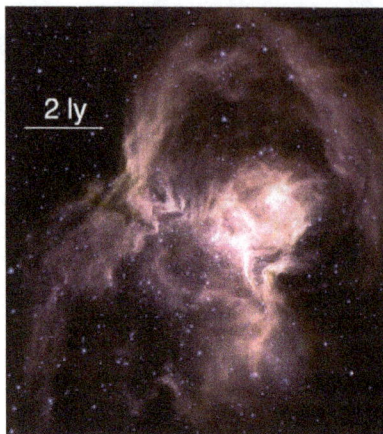

The Spitzer Space Telescope captures PAH spectral lines, producing this image of nebulosity in a stellar nursery.

PAHs may be abundant in the universe. They seem to have been formed as early as a couple of billion years after the Big Bang, and are associated with new stars and exoplanets. More than 20% of the carbon in the universe may be associated with PAHs. PAHs are considered possible starting material for the earliest forms of life. Light emitted by the Red Rectangle nebula and found spectral signatures that suggest the presence of anthracene and pyrene. This report was considered a controversial hypothesis that as nebulae of the same type as the Red Rectangle approach the ends of their lives, convection currents cause carbon and hydrogen in the nebulae's core to get caught in stellar winds, and radiate outward. As they cool, the atoms supposedly bond to each other in various ways and eventually form particles of a million or more atoms. Witt and his team inferred that PAHs —which may have been vital in the formation of early life on Earth— in a nebula, by necessity they must originate in nebulae.

Two extremely bright stars illuminate a mist of PAHs in this Spitzer image.

More recently, fullerenes (or "buckyballs"), have been detected in other nebulae. Fullerenes are also implicated in the origin of life; according to astronomer Letizia Stanghellini, "It's possible that buckyballs from outer space provided seeds for life on Earth." In September 2012, NASA scientists reported results of analog studies *in vitro* that PAHs, subjected to interstellar medium (ISM) conditions, are transformed, through hydrogenation, oxygenation, and hydroxylation, to more complex organics—"a step along the path toward amino acids and nucleotides, the raw materials of proteins and DNA, respectively". Further, as a result of these transformations, the PAHs lose their spectroscopic signature which could be one of the reasons "for the lack of PAH detection in interstellar ice grains, particularly the outer regions of cold, dense clouds or the upper molecular layers of protoplanetary disks."

PAHs have been detected in the upper atmosphere of Titan, the largest moon of the planet Saturn.

Dioxins and Dioxin-like Compounds

Dioxins and dioxin-like compounds (DLCs) are compounds that are highly toxic environmental persistent organic pollutants (POPs). They are mostly by-products of various industrial processes - or, in case of dioxin-like PCBs and PBBs, part of intentionally produced mixtures. They include:

- Polychlorinated dibenzo-*p*-dioxins (PCDDs), or simply dioxins. PCDDs are derivatives of dibenzo-*p*-dioxin. There are 75 PCDD congeners, differing in the number and location of chlorine atoms, and seven of them are especially toxic, the most dangerous being 2,3,7,8-Tetrachlorodibenzodioxin (TCDD)

- Polychlorinated dibenzofurans (PCDFs), or furans. PCDFs are derivatives of dibenzofuran. There are 135 isomers, ten have dioxin-like properties.

- Polychlorinated/polybrominated biphenyls (PCBs/PBBs), derived from biphenyl, of which twelve are "dioxin-like". Under certain conditions PCBs may form dibenzofurans/dioxins through partial oxidation.

- Finally, dioxin may refer to 1,4-Dioxin proper, the basic chemical unit of the more complex dioxins. This simple compound is not persistent and has no PCDD-like toxicity.

Because dioxins refer to such a broad class of compounds that vary widely in toxicity, the concept of toxic equivalence (TEQ) has been developed to facilitate risk assessment and regulatory control. Toxic equivalence factors (TEFs) exist for seven congeners of dioxins, ten furans and twelve PCBs. The reference congener is the most toxic dioxin 2,3,7,8-tetrachlorodibenzo-*p*-dioxin (TCDD) which per definition has a TEF of one.

In reference to their importance as environmental toxicants the term dioxins is used almost exclusively to refer to the sum of compounds (as TEQ) from the above groups which demonstrate the same specific toxic mode of action associated with TCDD. These include 17 PCDD/Fs and 12 PCBs. Incidents of contamination with PCBs are also often reported as dioxin contamination incidents since it is this toxic characteristic which is of most public and regulatory concern.

Toxicity

Mechanism of Toxicity

The toxic effects of dioxins are measured in fractional equivalencies of TCDD (2,3,7,8-tetrachlorodibenzo-p-dioxin), the most toxic and best studied member of its class. The toxicity is mediated through the interaction with a specific intracellular protein, the aryl hydrocarbon (AH) receptor, a transcriptional enhancer, affecting a number of other regulatory proteins. This receptor is a transcription factor which is involved in expression of many genes. TCDD binding to the AH receptor induces the cytochrome P450 1A class of enzymes which function to break down toxic compounds, e.g., carcinogenic polycyclic hydrocarbons such as benzo(a)pyrene (but making many of them more toxic in the process).

While the affinity of dioxins and related industrial toxicants to this receptor may not fully explain all their toxic effects including immunotoxicity, endocrine effects and tumor promotion, toxic responses appear to be typically dose-dependent within certain concentration ranges. A multiphasic dose-response relationship has also been reported, leading to uncertainty and debate about the true role of dioxins in cancer rates.

The endocrine disrupting activity of dioxins is thought to occur as a down-stream function of AH receptor activation, with thyroid status in particular being a sensitive marker of exposure. It is important to note that TCDD, along with the other PCDDs, PCDFs and dioxin-like coplanar PCBs are not direct agonists or antagonists of hormones, and are not active in assays which directly screen for these activities such as ER-CALUX and AR-CALUX. These compounds have also not been shown to have any direct mutagenic or genotoxic activity. Their main action in causing cancer is cancer promotion. A mixture of PCBs such as Aroclor may contain PCB compounds which are known estrogen agonists, but on the other hand are not classified as dioxin-like in terms of toxicity. Mutagenic effects have been established for some lower chlorinated chemicals such as 3-chlorodibenzofuran, which is neither persistent nor an AH receptor agonist.

Toxicity in Animals

The symptoms reported to be associated with dioxin toxicity in animal studies are incredibly wide ranging, both in the scope of the biological systems affected and in the range of dosage needed to bring these about. Acute effects of single high dose dioxin exposure include wasting syndrome, and typically a delayed death of the animal in 1 to 6 weeks. By far most toxicity studies have been performed using 2,3,7,8-tetrachlorodibenzo-*p*-dioxin.

The LD_{50} of TCDD varies wildly between species and even strains of the same species, with the most notable disparity being between the seemingly similar species of hamster and guinea pig. The oral LD_{50} for guinea pigs is as low as 0.5 to 2 µg/kg body weight, whereas the oral LD_{50} for hamsters can be as high as 1 to 5 mg/kg body weight. Even between different mouse or rat strains there may be tenfold to thousandfold differences in acute toxicity. Many pathological findings are seen in the liver, thymus and other organs.

Some chronic and sub-chronic exposures can be harmful at much lower levels, especially at particular developmental stages including foetal, neonatal and pubescent stages. Well established developmental effects are cleft palate, hydronephrosis, disturbances in tooth development and sexual development as well as endocrine effects.

Human Toxicity

Dioxins have been considered highly toxic and able to cause reproductive and developmental problems, damage the immune system, interfere with hormones and also cause cancer. This is based on animal studies. The best proven is chloracne. Even in poisonings with huge doses of TCDD, the only persistent effects after the initial malaise have been chloracne and amenorrhea. In occupational settings many symptoms have been seen, but exposures have always been to a multitude of chemicals including chlorophenols, chlorophenoxy acid herbicides, and solvents. Therefore, proof of dioxins as causative factors has been difficult. The suspected effects in adults are liver damage, and alterations in heme metabolism, serum lipid levels, thyroid functions, as well as diabetes and immunological effects.

In line with animal studies, developmental effects may be much more important than effects in adults. These include disturbances of tooth development, and of sexual development. An example of the variation in responses is clearly seen in a study following the Seveso disaster indicating that sperm count and motility were affected in different ways in exposed males, depending on whether they were exposed before, during or after puberty.

Intrauterine exposure to dioxins and dioxin-like compounds as an environmental toxin in pregnancy has subtle effects on the child later in life that include changes in liver function, thyroid hormone levels, white blood cell levels, and decreased performance in tests of learning and intelligence.

Exposure to dioxins can happen in a number of ways, most often as by-products of industrial waste. However, dioxins can result from natural processes including volcanic eruptions and forest fires, and manufacturing processes such as smelting, chlorine bleaching of paper pulp, and the creation of some herbicides and pesticides. Even at levels 100X lower than those associated with its cancer causing effects, the presence of dioxin can cause immune system damage, severe reproductive and developmental problems, and interference with regulatory hormones.

The Endometriosis Research Center (ERC) has testified before the California State Legislature concerning Assembly Bill 2820 [Cardoza, D-Merced] that, "feminine hygiene products (i.e. tampons) do indeed test positive for Dioxin. Dioxin, in turn, is a well-documented catalyst for Endometriosis - and the effects of Dioxin are cumulative; able to be measured as much as 20 or 30 years after exposure." The ERC also references an independent study that found, in an assessment of four brands of tampons and four brands of baby diapers, dioxins "were present at detectable concentrations in all samples." The presence of this toxin in tampons may be linked to endometriosis because dioxins last a long time in the body; they are chemically stable and can be absorbed by fat tissue, where they are then stored in the body. Their half-life in the body is estimated to be 7 to 11 years.

Carcinogenicity

Dioxins are well established carcinogens in animal studies, although the precise mechanistic role is not clear. Dioxins are not mutagenic or genotoxic. The United States Environmental Protection Agency has categorised dioxin, and the mixture of substances associated with sources of dioxin toxicity as a "likely human carcinogen". The International Agency for Research on Cancer has classified TCDD as a human carcinogen (class 1) on the basis of clear animal carcinogenicity and limited human data, but was not able to classify other dioxins. It is thought that the presence of dioxin can accelerate the formation of tumours and adversely affect the normal mechanisms for inhibiting tumour growth, without actually instigating the carcinogenic event.

As with all toxic endpoints of dioxin, a clear dose-response relationship is very difficult to establish. After accidental or high occupational exposures there is evidence on human carcinogenicity. There is much controversy especially on cancer risk at low population levels of dioxins. Among fishermen with high dioxin concentrations in their bodies, cancer deaths were decreased rather than increased. Some researchers have also proposed that dioxin induces cancer progression through a very different mitochondrial pathway.

Risk assessment

The uncertainty and variability in the dose-response relationship of dioxins in terms of their toxicity, as well as the ability of dioxins to bioaccumulate mean that the tolerable daily intake (TDI) of dioxin has been set very low, 1-4 pg/kg body weight per day, i.e. $7x10^{-11}$ to $2.8x10^{-10}$g per 70-kg person per day, to allow for this uncertainty and ensure public safety in all instances. Specifically, the TDI has been assessed based on the safety of children born to mothers exposed all their lifetime prior to pregnancy to such a daily intake of dioxins. It is likely that the TDI for other population groups could be somewhat higher. The most important cause for differences in different assessments is carcinogenicity. If the dose-response of TCDD in causing cancer is linear, it might be a true risk. If the dose-response is of a threshold-type or J-shape, there is little or no risk at the present concentrations. Understanding the mechanisms of toxicity better is hoped to increase the reliability of risk assessment.

Controversy

Greenpeace and some other environmental groups have called for the chlorine industry to be phased out. However, chlorine industry supporters say that "banning chlorine would mean that millions of people in the third world would die from want of disinfected water".(Although critics point out the existence of alternative water disinfection methods).

Sharon Beder and others have argued that the dioxin controversy has been very political and that large companies have tried to play down the seriousness of the problems of dioxin. The companies involved have often said that the campaign against dioxin is based on "fear and emotion" and not on science.

In 2008, Chile experienced a pork crisis caused by high dioxin concentrations in their pork exports. The contamination was found to be due to zinc oxide used in pork feed, and caused reputational and financial losses for the country, as well as leading to the introduction of new food safety regulations.

Human Intake and Levels

Most intake of dioxin-like chemicals is from food of animal origin: meat, dairy products, or fish predominate, depending on the country. The daily intake of dioxins and dioxin-like PCBs as TEQ is of the order of 100 pg/day, i.e. 1-2 pg/kg/day. In many countries both the absolute and relative significance of dairy products and meat have decreased due to strict emission controls, and brought about the decrease of total intake. E.g. in the United Kingdom the total intake of PCDD/F in 1982 was 239 pg/day and in 2001 only 21 pg/day (WHO-TEQ). Since the half-lives are very long (for e.g. TCDD 7–8 years), the body burden will increase almost over the whole lifetime. Therefore, the concentrations may increase five- to tenfold from age 20 to age 60. For the same reason, short term higher intake such as after food contamination incidents, is not crucial unless it is extremely high or lasts for several months or years.

The highest body burdens were found in Western Europe in the 1970s and early 1980s, and the trends have been similar in the U.S. The most useful measure of time trends is concentration in breast milk measured over decades. In many countries the concentrations have decreased to about one tenth of those in the 1970s, and the total TEQ concentrations are now of the order of 10-30 pg/g fat (please note the units, pg/g is the same as ng/kg, or the non-standard expression ppt used sometimes in America). The decrease is due to strict emission controls and also to the control of concentrations in food. In the U.S. young adult female population (age group 20-39), the concentration was 9.7 pg/g lipid in 2001-2002 (geometric mean).

Certain professions such as subsistence fishermen in some areas are exposed to exceptionally high amounts of dioxins and related substances. This along with high industrial exposures may be the most valuable source of information on the health risks of dioxins.

Uses

Dioxins have no common uses. They are manufactured on a small scale for chemical and toxicological research, but mostly exist as by-products of industrial processes such as bleaching paper pulp, pesticide manufacture, and combustion processes such as waste incineration. The defoliant Agent Orange contained dioxins. The production and use of dioxins was banned by the Stockholm Convention in 2001.

Sources

Environmental Sources

PCB-compounds, always containing low concentrations of dioxin-like PCBs and PCDFs, were synthesized for various technical purposes. They have entered the environment through accidents such as fires or leaks from transformers or heat exchangers, or from PCB-containing products in landfills or during incineration. Because PCBs are somewhat volatile, they have also been transported long distances by air leading to global distribution including the Arctic.

PCDD/F-compounds were never synthesized for any purpose, except for small quantities for scientific research. Small amounts of PCDD/Fs are formed whenever organics, oxygen and chlorine are available at suitable temperatures. This is augmented by metal catalysts such as copper. The optimal temperature range is 400 °C to 700 °C. This means that formation is highest when organic material is burned in less-than-optimal conditions such as open fires, building fires, domestic fireplaces, and poorly operated and/or designed solid waste incinerators. Historically, municipal and medical waste incineration was the most important source of PCDD/Fs.

Other sources of PCDD/F include:

- Metal smelting and refining

- Chlorine bleaching of pulp and paper - historically important source of PCDD/Fs to waterways.

- Synthesis side products of several chemicals, especially PCBs, chlorophenols, chlorophenoxy acid herbicides and hexachlorophene.

- Uncontrolled combustion, particularly open burning of waste ("backyard barrel burning"), accidental fires, wildfires.

- (Historical) Engines using leaded fuel, which contained the additives 1,2-Dichloroethane and 1,2-Dibromoethane.

In Waste Incineration

Improvements and changes have been made to nearly all industrial sources to reduce PCDD/F production. In waste incineration, large amounts of publicity and concern surrounded dioxin-like compounds during the 1980s-1990s and continues to pervade the public consciousness, especially when new incineration and waste-to-energy facilities are proposed. As a result of these concerns, incineration processes have been improved with increased combustion temperatures (over 1000 °C), better furnace control, and sufficient residence time allotted to ensure complete oxidation of organic compounds. Ideally, an incineration process oxidizes all carbon to CO_2 and converts all chlorine to HCl or inorganic chlorides prior to the gases passing through the temperature window of 700-400 °C where PCDD/F formation is possible. These substances cannot easily form organic compounds, and HCl is easily and safely neutralized in the scrubber while CO_2 is vented to the atmosphere. Inorganic chlorides are incorporated into the ash.

Scrubber and particulate removal systems manage to capture most of the PCDD/F which forms even in sophisticated incineration plants. These PCDD/Fs are generally not destroyed but moved into the fly ash. Catalytic systems have been designed which destroy vapor-phase PCDD/Fs at relatively low temperatures. This technology is often combined with the baghouse or SCR system at the tail end of an incineration plant.

European Union limits for concentration of dioxin-like compounds in the discharged flue gas is 0.1 ng/Nm³ TEQ.

Both in Europe and in U.S.A., the emissions have decreased dramatically since the 1980s, by even 90%. This has also led to decreases in human body burdens, which is neatly demonstrated by the decrease of dioxin concentrations in breast milk.

Open burning of waste (backyard barrel burning) has not decreased effectively, and in the U.S. it is now the most important source of dioxins. Total U.S. annual emissions decreased from 14 kg in 1987 to 1.4 kg in 2000. However, backyard barrel burning decreased only modestly from 0.6 kg to 0.5 kg, resulting in over one third of all dioxins in the year 2000 from backyard burning alone.

Low concentrations of dioxins have been found in some soils without any anthropogenic contamination. A puzzling case of milk contamination was detected in Germany.

The source was found to be kaolin added to animal feed. Dioxins have been repeatedly detected in clays from Europe and USA since 1996, with contamination of clay assumed to be the result of ancient forest fires or similar natural events with concentration of the PCDD/F during clay sedimentation.

Environmental Persistence and Bioaccumulation

All groups of dioxin-like compounds are persistent in the environment. Neither soil microbes nor animals are able to break down effectively the PCDD/Fs with lateral chlorines (positions 2,3,7, and 8). This causes very slow elimination. Ultraviolet light is able to slowly break down these compounds. Lipophilicity (tendency to seek for fat-like environments) and very poor water solubility make these compounds move from water environment to living organisms having lipid cell structures. This is called bio-accumulation. Increase in chlorination increases both stability and lipophilicity. The compounds with the very highest chlorine numbers (e.g. octachlorodibenzo-p-diox-in) are, however, so poorly soluble that this hinders their bioaccumulation. Bioaccu-mulation is followed by biomagnification. Lipid-soluble compounds are first accumu-lated to microscopic organisms such as phytoplankton (plankton of plant character, e.g. algae). Phytoplankton is consumed by animal plankton, this by invertebrates such as insects, these by small fish, and further by large fish and seals. At every stage or trophic level, the concentration is higher, because the persistent chemicals are not "burned off" when the higher organism uses the fat of the prey organism to produce energy.

Due to bioaccumulation and biomagnification, the species at the top of the trophic pyr-amid are most vulnerable to dioxin-like compounds. In Europe, the white-tailed eagle and some species of seals have approached extinction due to poisoning by persistent organic pollutants. Likewise, in America, the population of bald eagles declined be-cause of POPs causing thinning of eggs and other reproductive problems. Usually, the failure has been attributed mostly to DDT, but dioxins are also a possible cause of re-productive effects. Both in America and in Europe, many waterfowl have high concen-trations of dioxins, but usually not high enough to disturb their reproductive success. Due to supplementary winter feeding and other measures also, the white-tailed eagle is recovering. Also, ringed seals in the Baltic Sea are recovering.

Humans are also at the top of the trophic pyramid, particularly newborns. Exclusively breastfed newborns were estimated to be exposed to a total of 800 pg TEQ/day, leading to an estimated body weight-based dose of 242 pg TEQ/kg/day. Due to a multitude of food sources of adult humans exposure is much less averaging at 1 pg TEQ/kg-day, and dioxin concentrations in adults are much less at 10-100 pg/g, compared with 9000 to 340,000 pg/g (TEQ in lipid) in eagles or seals feeding almost exclusively on fish.

Because of different physicochemical properties, not all congeners of dioxin-like com-pounds find their routes to human beings equally well. Measured as TEQs, the dominant congeners in human tissues are 2,3,7,8-TCDD, 1,2,3,7,8-PeCDD, 1,2,3,6,7,8-Hx-CDD and 2,3,4,7,8-PeCDF. This is very different from most sources where hepta- and octa-congeners may predominate. The WHO panel re-evaluating the TEF values in 2005 expressed their concern that emissions should not be uncritically measured as TEQs, because all congeners are not equally important. They stated that "when a human risk assessment is to be done from abiotic matrices, factors such as fate, transport, and bioavailability from each matrix be specifically considered".

All POPs are poorly water-soluble, especially dioxins. Therefore, ground water contamination has not been a problem, even in cases of severe contamination due to the main chemicals such as chlorophenols. In surface waters, dioxins are bound to organic and inorganic particles.

Fate of Dioxins in Human Body

The same features causing persistence of dioxins in the environment, also cause very slow elimination in humans and animals. Because of low water solubility, kidneys are not able to secrete them in urine as such. They should be metabolised to more water-soluble metabolites, but also metabolism especially in humans is extremely slow. This results in biological half-lives of several years for all dioxins. That of TCDD is estimated to be 7 to 8 years, and for other PCDD/Fs from 1.4 to 13 years, PCDFs on average slightly shorter than PCDDs.

Dioxins are absorbed well from the digestive tract, if they are dissolved in fats or oils (e.g. in fish or meat). On the other hand, dioxins tend to adsorb tightly to soil particles, and absorption may be quite low: 13.8% of the given dose of TEQs in contaminated soil was absorbed.

In mammalian organisms, dioxins are found mostly in fat. Concentrations in fat seem to be relatively similar, be it serum fat, adipose tissue fat, or milk fat. This permits measuring dioxin burden by analysing breast milk. Initially, however, at least in laboratory animals, after a single dose, high concentrations are found in the liver, but in a few days, adipose tissue will predominate. In rat liver, however, high doses cause induction of CYP1A2 enzyme, and this binds dioxins. Thus, depending on the dose, the ratio of fat and liver tissue concentrations may vary considerably in rodents.

Sources of Human Exposure

The most important source of human exposure is fatty food of animal origin, and breast milk. There is much variation between different coun-tries as to the most important items. In U.S. and Central Europe, milk, dairy products and meat have been by far the most important sources. In some countries, notably in Finland and to some extent in Sweden, fish is important due to contaminated Baltic fish and very low intake from any other sources. In most countries, a significant decrease of dioxin intake has occurred due to stricter controls during the last 20 years.

Historically occupational exposure to dioxins has been a major problem. Dioxins are formed as important toxic side products in the production of PCBs, chlorophenols, chlorophenoxy acid herbicides, and other chlorinated organic chemicals. This caused very high exposures to workers in poorly controlled hygienic conditions. Many workers had chloracne. In a NIOSH study in the U.S., the average concentration of TCDD in exposed persons was 233 ng/kg (in serum lipid) while it was 7 ng/kg in unexposed workers, even though the exposure had been 15–37 years earlier. This indicates a huge previous exposure. In fact the exact back-calculation is debated, and the concentrations may have been even several times higher than originally estimated.

Handling and spraying of chlorophenoxy acid herbicides may also cause quite high exposures, as clearly demonstrated by the users of Agent Orange in the Malayan Emergency and in the Vietnam War. The highest concentrations were detected in nonflying enlisted personnel (e.g. filling the tanks of planes), although the variation was huge, 0 to 618 ng/kg TCDD (mean 23.6 ng/kg). Other occupational exposures (working at paper and pulp mills, steel mills and incinerators) have been remarkably lower.

Accidental exposures have been huge in some cases. The highest concentrations in people after the Seveso accident were 56,000 ng/kg, and the highest exposure ever recorded was found in Austria in 1998, 144,000 ng/kg. This is equivalent to a dose of 20 to 30 µg/kg TCDD, a dose that would be lethal to guinea pigs and some rat strains.

Exposure from contaminated soil is possible when dioxins are blown up in dust, or children eat dirt. Inhalation was clearly demonstrated in Missouri in the 1970s, when waste oils were used as dust suppressant in horse arenas. Many horses and other animals were killed due to poisoning. Dioxins are neither volatile nor water-soluble, and therefore exposure of human beings depends on direct eating of soil or production of dust which carries the chemical. Contamination of ground water or breathing vapour of the chemical are not likely to cause a significant exposure. Currently, in the US, there are 126 Superfund sites with a completed exposure pathway contaminated with dioxins.

Further, PCBs are known to pass through treatment plants and accumulate in sludge which is used on farm fields in certain countries. In 2011 in South Carolina, SCDHEC enacted emergency sludge regulations after PCBs were found to have been discharged to a waste treatment plant.

PCBs are also known to flush from industry and land (aka sludge fields) to contaminate fish, as they have up and down the Catawba River in North and South Carolina. State authorities have posted fish consumption advisories due to accumulation of PCBs in fish tissue.

TEF Values

All dioxin-like compounds share a common mechanism of action via the aryl hydrocarbon receptor (AHR), but their potencies are very different. This means that similar effects are caused by all of them, but much larger doses of some of them are needed than of TCDD. Binding to the AHR as well as persistence in the environment and in the organism depends on the presence of so-called "lateral chlorines", in case of dioxins and furans, chlorine substitutes in positions 2,3,7, and 8. Each additional non-lateral chlorine decreases the potency, but qualitatively the effects remain similar. Therefore, a simple sum of different dioxin congeners is not a meaningful measure of toxicity. To compare the toxicities of various congeners and to render it possible to make a toxicologically meaningful sum of a mixture, a toxicity equivalency (TEQ) concept was created.

Each congener has been given a toxicity equivalence factor (TEF). This indicates its relative toxicity as compared with TCDD. Most TEFs have been extracted from *in vivo* toxicity data on animals, but if these are missing (e.g. in case of some PCBs), less reliable *in vitro* data have been used. After multiplying the actual amount or concentration of a congener by its TEF, the product is the virtual amount or concentration of TCDD having effects of the same magnitude as the compound in question. This multiplication is done for all compounds in a mixture, and these "equivalents of TCDD" can then simply be added, resulting in TEQ, the amount or concentration of TCDD toxicologically equivalent to the mixture.

The TEQ conversion makes it possible to use all studies on the best studied TCDD to assess the toxicity of a mixture. This resembles the common measure of all alcoholic drinks: beer, wine and whiskey can be added together as absolute alcohol, and this sum gives the toxicologically meaningful measure of the total impact.

The TEQ only applies to dioxin-like effects mediated by the AHR. Some toxic effects (especially of PCBs) may be independent of the AHR, and those are not taken into account by using TEQs.

TEFs are also approximations with certain amount of scientific judgement rather than scientific facts. Therefore, they may be re-evaluated from time to time. There have been several TEF versions since the 1980s. The most recent re-assessment was by an expert group of the World Health organization in 2005.

References

- Weil, ED; Levchik, SV (2015). Flame Retardants for Plastics and Textiles: Practical Applications. Munich: Carl Hanser Verlag. p. 97. ISBN 1569905789.

- Robertson, edited by Larry W.; Hansen, Larry G. (2001). PCBs : recent advances in environmental toxicology and health effects. Lexington, Ky.: University Press of Kentucky. p. 11. ISBN 0813122260.

- Kimbrough, R. D.; Jensen, A. A. (2012-12-02). Halogenated Biphenyls, Terphenyls, Naphthalenes, Dibenzodioxins and Related Products. Elsevier. p. 24. ISBN 9780444598929.

- Identifying PCB-Containing Capacitors (PDF). Australian and New Zealand Environment and Conservation Council. 1997. pp. 4–5. ISBN 0-642-54507-3. Retrieved 2007-07-07.

- Steingraber, Sandra (2001). "12". Having Faith : an ecologist's journey to motherhood (Berkley trade pbk. ed.). New York: Berkley. ISBN 0425189996.

- Bull J, Farrand, J Jr (1987). Audubon Society Field Guide to North American Birds: Eastern Region. New York: Alfred A. Knopf. pp. 468–9. ISBN 0-394-41405-5

- Przyrembel H, Heinrich-Hirsch B, Vieth B; Heinrich-Hirsch; Vieth (2000). "Exposition to and health effects of residues in human milk". Adv. Exp. Med. Biol. Advances in Experimental Medicine and Biology. 478: 307–25. doi:10.1007/0-306-46830-1_27. ISBN 0-306-46405-5.

- Harvey, R.G. (1998). "Environmental Chemistry of PAHs". PAHs and Related Compounds: Chemistry. The Handbook of Environmental Chemistry. Springer. pp. 1–54. ISBN 9783540496977.

Agricultural Pollutants

Pollutants that contaminate agricultural practices can be classified into insects, DDT, fungicides and pesticides. Insecticides are substances used in extermination of insects and fungicides are used to kill fungi or fungal spores. The following section is a compilation of the various branches of agricultural pollutants that form an integral part of the broader subject matter.

Insecticide

An insecticide is a substance used to kill insects. They include ovicides and larvicides used against insect eggs and larvae, respectively. Insecticides are used in agriculture, medicine, industry and by consumers. Insecticides are claimed to be a major factor behind the increase in agricultural 20th century's productivity. Nearly all insecticides have the potential to significantly alter ecosystems; many are toxic to humans; some concentrate along the food chain.

FLIT manual spray pump for insecticides from 1928

Insecticides can be classified in two major groups: systemic insecticides, which have residual or long term activity; and contact insecticides, which have no residual activity.

Furthermore, one can distinguish three types of insecticide. 1. Natural insecticides, such as nicotine, pyrethrum and neem extracts, made by plants as defenses against insects. 2. Inorganic insecticides, which are metals. 3. Organic insecticides, which are organic chemical compounds, mostly working by contact.

The mode of action describes how the pesticide kills or inactivates a pest. It provides another way of classifying insecticides. Mode of action is important in understanding whether an insecticide will be toxic to unrelated species, such as fish, birds and mammals.

Insecticides are distinct from insect repellents, which do not kill.

Type of Activity

Systemic insecticides become incorporated and distributed systemically throughout the whole plant. When insects feed on the plant, they ingest the insecticide. Systemic insecticides produced by transgenic plants are called plant-incorporated protectants (PIPs). For instance, a gene that codes for a specific Bacillus thuringiensis biocidal protein was introduced into corn and other species. The plant manufactures the protein, which kills the insect when consumed.

Contact insecticides are toxic to insects upon direct contact. These can be inorganic insecticides, which are metals and include arsenates, copper and fluorine compounds, which are less commonly used, and the commonly used sulfur. Contact insecticides can be organic insecticides, i.e. organic chemical compounds, synthetically produced, and comprising the largest numbers of pesticides used today. Or they can be natural compounds like pyrethrum, neem oil etc. Contact insecticides usually have no residual activity.

Efficacy can be related to the quality of pesticide application, with small droplets, such as aerosols often improving performance.

Biological Pesticides

Many organic compounds are produced by plants for the purpose of defending the host plant from predation. A trivial case is tree rosin, which is a natural insecticide. Specific, the production of oleoresin by conifer species is a component of the defense response against insect attack and fungal pathogen infection. Many fragrances, e.g. oil of wintergreen, are in fact antifeedants.

Four extracts of plants are in commercial use: pyrethrum, rotenone, neem oil, and various essential oils

Other Biological Approaches

Plant-incorporated Protectants

Transgenic crops that act as insecticides began in 1996 with a genetically modified potato that produced the Cry protein, derived from the bacterium Bacillus thuringiensis, which is toxic to beetle larvae such as the Colorado potato beetle. The technique has been expanded to include the use of RNA interference RNAi that fatally silences crucial insect genes. RNAi likely evolved as a defense against viruses. Midgut cells in many larvae take up the molecules and help spread the signal. The technology can target only insects that have the silenced sequence, as was demonstrated when a particular RNAi affected only one of four fruit fly species. The technique is expected to replace many other insecticides, which are losing effectiveness due to the spread of pesticide resistance.

Enzymes

Many plants exude substances to repel insects. Premier examples are substances activated by the enzyme myrosinase. This enzyme converts glucosinolates to various compounds that are toxic to herbivorous insects. One product of this enzyme is allyl isothiocyanate, the pungent ingredient in horseradish sauces.

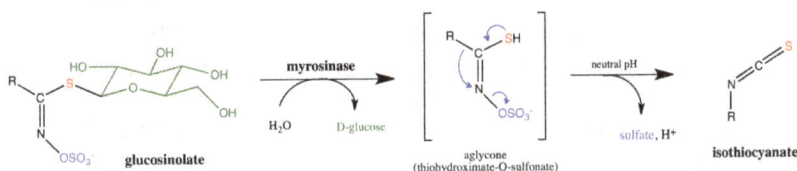

Biosynthesis of antifeedants by the action of myrosinase.

The myrosinase is released only upon crushing the flesh of horseradish. Since allyl isothiocyanate is harmful to the plant as well as the insect, it is stored in the harmless form of the glucosinolate, separate from the myrosinase enzyme.

Bacterial

Bacillus thuringiensis is a bacterial disease that affects Lepidopterans and some other insects. Toxins produced by strains of this bacterium are used as a larvicide against caterpillars, beetles, and mosquitoes. Toxins from *Saccharopolyspora spinosa* are isolated from fermentations and sold as Spinosad. Because these toxins have little effect on other organisms, they are considered more environmentally friendly than synthetic pesticides. The toxin from *B. thuringiensis* (Bt toxin) has been incorporated directly into plants through the use of genetic engineering. Other biological insecticides include products based on entomopathogenic fungi (e.g., *Beauveria bassiana*, *Metarhizium anisopliae*), nematodes (e.g., *Steinernema feltiae*) and viruses (e.g., *Cydia pomonella* granulovirus).

Synthetic Insecticide

A major emphasis of organic chemistry is the development of chemical tools to enhance agricultural productivity. Insecticides represent a major area of emphasis. Many of the major insecticides are inspired by biological analogues. Many others are completely alien to nature.

Organochlorides

The best known organochloride, DDT, was created by Swiss scientist Paul Müller. For this discovery, he was awarded the 1948 Nobel Prize for Physiology or Medicine. DDT was introduced in 1944. It functions by opening sodium channels in the insect's nerve cells. The contemporaneous rise of the chemical industry facilitated large-scale production of DDT and related chlorinated hydrocarbons.

Organophosphates and Carbamates

Organophosphates are another large class of contact insecticides. These also target the insect's nervous system. Organophosphates interfere with the enzymes acetylcholinesterase and other cholinesterases, disrupting nerve impulses and killing or disabling the insect. Organophosphate insecticides and chemical warfare nerve agents (such as sarin, tabun, soman, and VX) work in the same way. Organophosphates have a cumulative toxic effect to wildlife, so multiple exposures to the chemicals amplifies the toxicity. In the US, organophosphate use declined with the rise of substitutes.

Carbamate insecticides have similar mechanisms to organophosphates, but have a much shorter duration of action and are somewhat less toxic.

Pyrethroids

Pyrethroid pesticides mimic the insecticidal activity of the natural compound pyrethrum, the biopesticide found in pyrethrins. These compounds are nonpersistent sodium channel modulators and are less toxic than organophosphates and carbamates. Compounds in this group are often applied against household pests.

Neonicotinoids

Neonicotinoids are synthetic analogues of the natural insecticide nicotine (with much lower acute mammalian toxicity and greater field persistence). These chemicals are acetylcholine receptor agonists. They are broad-spectrum systemic insecticides, with rapid action (minutes-hours). They are applied as sprays, drenches, seed and soil treatments. Treated insects exhibit leg tremors, rapid wing motion, stylet withdrawal (aphids), disoriented movement, paralysis and death. Imidacloprid may be the most common. It has recently come under scrutiny for allegedly pernicious effects on honeybees and its potential to increase the susceptibility of rice to planthopper attacks.

Ryanoids

Ryanoids are synthetic analogues with the same mode of action as ryanodine, a naturally occurring insecticide extracted from *Ryania speciosa* (Flacourtiaceae). They bind to calcium channels in cardiac and skeletal muscle, blocking nerve transmission. Only one such insecticide is currently registered, Rynaxypyr, generic name chlorantraniliprole.

Insect Growth Regulators

Insect growth regulator (IGR) is a term coined to include insect hormone mimics and an earlier class of chemicals, the benzoylphenyl ureas, which inhibit chitin(exoskeleton) biosynthesis in insects. Diflubenzuron is a member of the latter class, used primarily to control caterpillars that are pests. The most successful insecticides in this class are the juvenoids (juvenile hormone analogues). Of these, methoprene is most widely used. It has no observable acute toxicity in rats and is approved by World Health Organization (WHO) for use in drinking water cisterns to combat malaria. Most of its uses are to combat insects where the adult is the pest, including mosquitoes, several fly species, and fleas. Two very similar products, hydroprene and kinoprene, are used for controlling species such as cockroaches and white flies. Methoprene was registered with the EPA in 1975. Virtually no reports of resistance have been filed. A more recent type of IGR is the ecdysone agonist tebufenozide (MIMIC), which is used in forestry and other applications for control of caterpillars, which are far more sensitive to its hormonal effects than other insect orders.

Environmental Effects

Effects on Nontarget Species

Some insecticides kill or harm other creatures in addition to those they are intended to kill. For example, birds may be poisoned when they eat food that was recently sprayed with insecticides or when they mistake an insecticide granule on the ground for food and eat it.

Sprayed insecticide may drift from the area to which it is applied and into wildlife areas, especially when it is sprayed aerially.

DDT

The development of DDT was motivated by desire to replace more dangerous or less effective alternatives. DDT was introduced to replace lead and arsenic-based compounds, which were in widespread use in the early 1940s.

DDT was brought to public attention by Rachel Carson's book *Silent Spring*. One side-effect of DDT is to reduce the thickness of shells on the eggs of predatory birds. The shells sometimes become too thin to be viable, reducing bird populations. This occurs with DDT and related compounds due to the process of bioaccumulation, wherein the chemical, due to its stability and fat solubility, accumulates in organisms' fatty tissues. Also, DDT may biomagnify, which causes progressively higher concentrations in the body fat of animals farther up the food chain. The near-worldwide ban on agricultural use of DDT and related chemicals has allowed some of these birds, such as the peregrine falcon, to recover in recent years. A number of organochlorine pesticides have been banned from most uses worldwide. Globally they are controlled via the Stockholm

Convention on persistent organic pollutants. These include: aldrin, chlordane, DDT, dieldrin, endrin, heptachlor, mirex and toxaphene.

Pollinator Decline

Insecticides can kill bees and may be a cause of pollinator decline, the loss of bees that pollinate plants, and colony collapse disorder (CCD), in which worker bees from a beehive or Western honey bee colony abruptly disappear. Loss of pollinators means a reduction in crop yields. Sublethal doses of insecticides (i.e. imidacloprid and other neonicotinoids) affect bee foraging behavior. However, research into the causes of CCD was inconclusive as of June 2007.

Organochloride Insecticides

Aldrin

Aldrin is an organochlorine insecticide that was widely used until the 1970s, when it was banned in most countries. It is a colourless solid. Before the ban, it was heavily used as a pesticide to treat seed and soil. Aldrin and related "cyclodiene" pesticides (a term for pesticides derived from Hexachlorocyclopentadiene) became notorious as persistent organic pollutants.

Production

Aldrin is produced by combining hexachlorocyclopentadiene with norbornadiene in a Diels-Alder reaction to give the adduct.

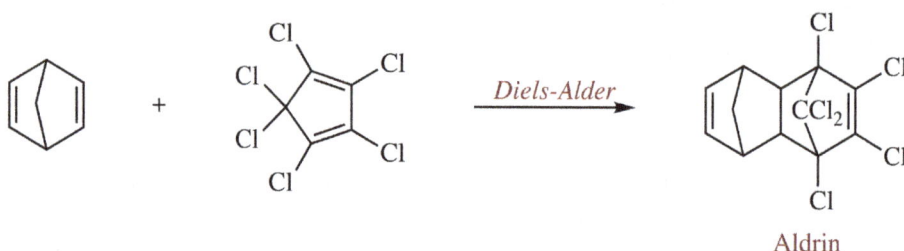

Synthesis of aldrin via a Diels-Alder reaction

Similarly, an isomer of aldrin, known as isodrin, is produced by reaction of hexachloronobornadiene with cyclopentadiene.

Aldrin is named after the German chemist Kurt Alder, one of the coinventors of this kind of reaction. An estimated 270 million kilograms of aldrin and related cyclodiene pesticides were produced between 1946 and 1976.

In soil, on plant surfaces, or in the digestive tracts of insects, aldrin oxidizes to the epoxide dieldrin, which is more strongly insecticidal.

Environmental Impact and Regulation

Like related polychlorinated pesticides, aldrin is highly lipophilic. Its solubility in water is only 0.027 mg/L, which exacerbates its persistence in the environment. It was banned by the Stockholm Convention on Persistent Organic Pollutants. In the U.S., aldrin was cancelled in 1974. The substance is banned from use for plant protection by the EU.

Safety and Environmental Aspects

Aldrin has rat LD_{50} of 39 to 60 mg/kg (oral in rats). For fish however, it is extremely toxic, with an LC50 of 0.006 – 0.01 for trout and bluegill.

In the US, aldrin is considered a potential occupational carcinogen by the Occupational Safety and Health Administration and the National Institute for Occupational Safety and Health; these agencies have set an occupational exposure limit for dermal exposures at 0.25 mg/m³ over an eight-hour time-weighted average. Further, an IDLH limit has been set at 25 mg/m³, based on acute toxicity data in humans to which subjects reacted with convulsions within 20 minutes of exposure.

It is classified as an extremely hazardous substance in the United States as defined in Section 302 of the U.S. Emergency Planning and Community Right-to-Know Act (42 U.S.C. 11002), and is subject to strict reporting requirements by facilities which produce, store, or use it in significant quantities.

Dieldrin

Dieldrin is an organochloride originally produced in 1948 by J. Hyman & Co, Denver, as an insecticide. Dieldrin is closely related to aldrin, which reacts further to form dieldrin. Aldrin is not toxic to insects; it is oxidized in the insect to form dieldrin which is the active compound. Both dieldrin and aldrin are named after the Diels-Alder reaction which is used to form aldrin from a mixture of norbornadiene and hexachlorocyclopentadiene.

Originally developed in the 1940s as an alternative to DDT, dieldrin proved to be a highly effective insecticide and was very widely used during the 1950s to early 1970s. Endrin is a stereoisomer of dieldrin.

However, it is an extremely persistent organic pollutant; it does not easily break down. Furthermore it tends to biomagnify as it is passed along the food chain. Long-term exposure has proven toxic to a very wide range of animals including humans, far greater than to the original insect targets. For this reason it is now banned in most of the world.

It has been linked to health problems such as Parkinson's, breast cancer, and immune, reproductive, and nervous system damage. It is also an endocrine disruptor, acting as an estrogen and antiandrogen, and can adversely affect testicular descent in the fetus if a pregnant woman is exposed to it.

Synthesis

Dieldrin can be formed from the synthesis of hexachloro-1,3-cyclopentadiene with nor-bornadiene in a Diels-Alder reaction, followed by epoxidation of the norbornene ring.

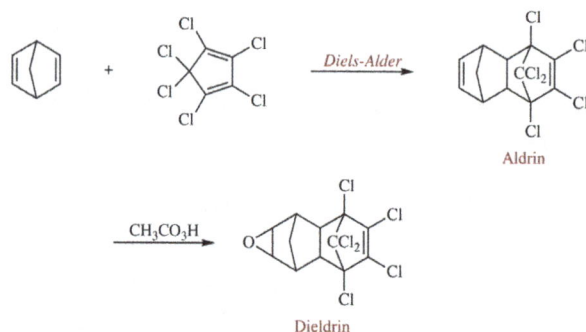

Synthesis of Dieldrin via a Diels-Alder reaction

Legislation and History

The chemicals dieldrin and aldrin were widely applied in agricultural areas throughout the world. Both are toxic and bioaccumulative. Aldrin does break down to dieldrin in living systems, but dieldrin is known to resist bacterial and chemical breakdown processes in the environment.

Aldrin was used to control soil pests (namely termites) on corn and potato crops. Dieldrin was an insecticide used on fruit, soil, and seed. It persists in the soil with a half-life of five years at temperate latitudes. Both aldrin and dieldrin may be volatilized from sediment and redistributed by air currents, contaminating areas far from their sources. They have been measured in Arctic wildlife, suggesting long range transport from southern agricultural regions.

Both aldrin and dieldrin have been banned in most developed countries, but aldrin is still used as a termiticide in Malaysia, Thailand, Venezuela and parts of Africa. In Canada, their sale was restricted in the mid-1970s, with the last registered use of the compounds in Canada being withdrawn in 1984.

IPCS quotes the World Health Organization as stating dieldrin is prohibited for use in agriculture in, among others, Brazil, Ecuador, Finland, the German Democratic Republic, Singapore, Sweden, Yugoslavia, and the USSR. The European Community legislation prohibits the marketing of phytopharmaceutical products containing dieldrin. In Argentina, Canada, Chile, the Federal Republic of Germany, Hungary, and the USA, its use is prohibited, with some exceptions. The use of dieldrin is restricted in India, Mauritius, Togo, and the United Kingdom. Its use in industry is prohibited in Switzerland

and its manufacture and use in Japan is under government control. In Finland, the only accepted use for dieldrin is as a termiticide in one glue mixture for exported plywood. India requires registration and licences for all importation, manufacture, sale, or storage.

Australia

Organochlorines and other chemicals were originally developed in the 1930s for use as insecticides and pesticides. DDT became famous worldwide in 1939 after its use in overcoming a typhus infestation in Naples. The use of organochlorines increased during the 1950s and peaked in the 1970s. Their use in Australia was dramatically lowered between the mid 1970s and the early 1980s. The first restrictions on the use of dieldrin and related chemicals in Australia were introduced in 1961-2, with registration required for their use on produce animals, such as cattle and chickens. This coincided with increasing concerns worldwide about the long-term effects of persistent pesticides. The publication of *Silent Spring* (a widely read and highly influential popular account of the environmental and health effects of pesticides) by Rachel Carson in 1962 was a key driving force in raising this concern. The phase-out process was driven by government bans and deregistration, in turn promoted by changing public perceptions that food containing residues of these chemicals was less acceptable and possibly hazardous to health.

Throughout this time, continuous pressure was maintained by relevant committees, for example the Technical Committee on Agricultural Chemicals (TCAC), to reduce approved organochlorine use. By 1981, the use of dieldrin worldwide was limited to sugarcane and bananas, and these uses were deregistered by 1985. In 1987, a nationwide recall system was put into place, and in December of that year, the government prohibited all imports of these chemicals into Australia without express ministerial approval. In 1994, the National Registration Authority for Agricultural and Veterinary Chemicals published a use of organochlorines in termite control, recommending the phase-out of organochlorines used in termite control upon development of viable alternatives. The same year, the Agriculture and Resource Management Council of Australia and New Zealand decided to phase out remaining organochlorine uses by 30 June 1995, with the exception of the Northern Territory. In November 1997, the use of all organochlorines other than mirex was phased out in Australia. Remaining stocks of mirex are to be used only for contained baits for termites in plantations of young trees in the Northern Territory until stocks run out, which is expected in the near future.

The recognition of negative impacts on health has stimulated the implementation of multiple legislative policies in regards to the use and disposal of organochlorine pesticides. For example, the Environment Protection (Marine) Policy 1994 became operational in May 1995 in South Australia. It dictated the acceptable concentration of toxicants such as dieldrin in marine waters and the manner in which these levels must be tested and tried.

Momentum against organochlorine and similar molecules continued to grow internationally, leading, to negotiations which matured as the Stockholm Convention on the use of persistent organic pollutants (POPs). POPs are defined as hazardous and environmentally persistent substances which can be transported between countries by the earth's oceans and atmosphere.

Most POPs (including dieldrin) bioaccumulate in the fatty tissues of humans and other animals. The Stockholm Convention banned 12 POPs, nicknamed "the dirty dozen". (These include: aldicarb, toxaphene, chlordane and heptachlor, chlordimeform, chlorobenzilate, DBCP, DDT, "drins" (aldrin, dieldrin and endrin), EDB, HCH and lindane, paraquat, parathion and methyl parathion, pentachlorophenol, and 2,4,5-T.) This took force on 17 May 2004. Australia ratified the Convention only three days later and became a party to it in August that year.

Well before this, Australia had been well advanced in meeting the measures agreed upon under the Convention. Production, import and use of aldrin, chlordane, DDT, dieldrin, hexachlorobenzene (HCB), heptachlor, endrin, and toxaphene are not permitted in Australia. Production and importation of polychlorinated biphenyls (PCBs) are not permitted in Australia, with the phase-out of existing PCBs being managed under the National Strategy for the Management of Scheduled Waste. This strategy also addresses how Australia will manage HCB waste and organochlorine pesticides.

Legislation in Australia on the import, use and disposal of dieldrin and other organochlorines has been extensive and covers mainly environmental and potential health impacts on the population.

Endrin

Endrin is an organochloride with the chemical formula $C_{12}H_8Cl_6O$ that was first produced in 1950 by Shell and Velsicol Chemical Corporation. It was primarily used as an insecticide, as well as a rodenticide and piscicide. It is a colourless, odorless solid, although commercial samples are often off-white. Endrin was manufactured as an emulsifiable solution known commercially as Endrex. The compound became infamous as a persistent organic pollutant and for this reason it is banned in many countries.

In the environment endrin exists as either endrin aldehyde or endrin ketone and can be found mainly in bottom sediments of bodies of water. Exposure to endrin can occur by inhalation, ingestion of substances containing the compound, or skin contact. Upon entering the body, it can be stored in body fats and can act as a neurotoxin on the central nervous system, which can cause convulsions, seizures, or even death.

Although endrin is not currently classified as a mutagen, nor as a human carcinogen, it is still a toxic chemical in other ways with detrimental effects. Due to these toxic effects, the manufacturers cancelled all use of endrin in the United States by 1991. Food import concerns have been raised because some countries may have still been using endrin as a pesticide.

History

J. Hyman & Company first developed endrin in 1950. Shell International was licensed in the United States and in the Netherlands to produce it. Velsicol was the other producer in the Netherlands. Endrin was used globally until the early 1970s. Due to its toxicity, it was banned or severely restricted in many countries. In 1982, Shell discontinued its manufacturing.

In 1962, an estimated 2.3-4.5 million kilograms of endrin were sold by Shell in the USA. In 1970, Japan imported 72,000 kilograms of endrin. From 1963 until 1972, Bali used 171 to 10,700 kilograms of endrin annually for the production of rice paddies until endrin use was discontinued in 1972. Taiwan reported to show higher levels of organochlorine pesticides including endrin in soil samples of paddy fields, compared to other Asian countries such as Thailand and Vietnam. During the 1950s-1970s over two million kilograms of organochlorine pesticides were estimated of having been be released into the environment per year. Endrin was banned in the United States on October 10, 1984. Taiwan banned endrin's use as a pesticide in 1971 and regulated it as a toxic chemical in 1989.

In May 2004, the Stockholm Convention on Persistent Organic Pollutants came into effect and listed endrin as one of the 12 initial persistent organic pollutants (POPs) that have been causing adverse effects on humans and the environment. The convention requires the participating parties to take measures to eliminate or restrict the production of POPs.

Production

The synthesis of endrin begins with the condensation of hexachlorocyclopentadiene with vinyl chloride. The product is then dehydrochlorinated. Following reaction with cyclopentadiene, isodrin is formed. Epoxide formation by adding either peracetic acid or perbenzoic acid to the isodrin is the final step in synthesizing endrin.

Endrin is a stereoisomer of dieldrin with comparable properties, though endrin degrades more easily

Use

Endrin was formulated as emulsifiable concentrates (ECs), wettable powders (WPs), granules, field strength dusts (FSDs), and pastes. The product could then be applied by aircraft or by handheld sprayers in its various formulations.

Endrin has been used primarily as an agricultural insecticide on tobacco, apple trees, cotton, sugar cane, rice, cereal, and grains. It is effective against a variety of species, including cotton bollworms, corn borers, cut worms and grass hoppers. In addition, endrin has been employed as a rodenticide and avicide. In Malaysia, fish farms used a solution of endrin as a piscicide to rid mine pools and fish ponds of all fish prior to restocking.

A study conducted from 1981 to 1983 in the US aimed to determine endrin's effects on non-target organisms when applied as a rodenticide in orchards. Most wildlife in and around the orchard was found to have endrin exposure, with endrin toxicity accounting for more than 24% of bird deaths recorded. Endrin was eventually banned in the US on October 10, 1984.

Health Effects

Exposure and Metabolism

Exposure to endrin can occur by inhalation, ingestion of substances containing the compound, or by skin contact. In addition to inhalation and skin contact, infants can be exposed by ingesting the breast milk of an exposed woman. *In utero*, fetuses are exposed by way of the placenta if the mother has been exposed.

Upon entering the body, endrin metabolizes into *anti*-12-hydroxyendrin and other metabolites, which can be expelled in the urine and feces. Both *anti*-12-hydroxyendrin and its metabolite, 12-ketoendrin, are likely responsible for the toxicity of endrin. The rapid metabolism of endrin into these metabolites makes detection of endrin itself difficult unless exposure is very high.

Neurological Effects

Symptoms of endrin poisoning include headache, dizziness, nervousness, confusion, nausea, vomiting, and convulsions. Acute endrin poisoning in humans affects primarily the central nervous system. There, it can act as a neurotoxin that blocks the activity of inhibitory neurotransmitters. In cases of acute exposure, this may result in seizures, or even death. Because endrin can be stored in body fats, acute endrin poisoning can lead to recurrent seizures when stressors induce the release of endrin back into the body, even months after the initial exposure is terminated.

People occupationally exposed to endrin may experience abnormal EEG readings even if they exhibit none of the clinical symptoms, possibly due to injury to the brain stem. These readings show bilateral synchronous theta waves with synchronous spike-and-wave complexes. EEG readings can take up to one month to return to normal.

Developmental Effects

Though endrin exposure has not been found to adversely affect fertility in mammals, an increase in fetal mortality has been observed in mice, rats, and mallard ducks. In those animals that have survived gestation, developmental abnormalities have been observed, particularly in rodents whose mothers were exposed to endrin early in pregnancy. In hamsters, the number of cases of fused ribs, cleft palate, open eyes, webbed feet, and meningoencephaloceles have increased. Along with open eyes and cleft palate, mice have developed with fused ribs and exencephaly. Skeletal abnormalities in rodents have also been reported.

Other Effects

Higher doses of endrin have been found to cause the following in rodents: renal tubular necrosis; inflammation of the liver, fatty liver, and liver necrosis; possible kidney degradation; and a decrease in body weight and body weight gain.

Endrin is very toxic to aquatic organisms, namely fish, aquatic invertebrates, and phytoplankton. It was found to remain in the tissues of infected fish for up to one month.

1984 Poisoning Outbreak in Pakistan

From July 14 to September 26, 1984, an outbreak of endrin poisoning occurred in 21 villages in and around Talagang, a subdistrict of the Punjab province of Pakistan. Eighty percent of the 194 known cases were children under the age of 15. Poisoned individuals had seizures along with vomiting, pulmonary congestion, and hypoxia, leaving 19 people dead. Some individuals had low grade fevers (37.8 °C/100 °F, axillary) following seizures. The more seriously affected had less vomiting, but higher temperatures than people who were less affected. Most patients could be controlled in under two hours using diazepam, phenobarbital, and atropine, though the more seriously affected patients required general anesthesia. Recovery took up to two days. Following treatment, patients reported not remembering their seizures. The outbreak affected both men and women equally.

Based on the demographics of the affected individuals and their area of residence, the outbreak was likely caused by endrin contamination of food. As members of these villages rarely had contact with one another, investigators determined that contaminated sugar shipped to the villages was the most probable cause, though no credible evidence was found to support this. Around this time, endrin was being used by cotton and sugar cane farmers in the Punjab region. A number of truck drivers stated that they had used the same trucks to deliver endrin to farmers and to pick up crops for Talagang, possibly leading to contamination.

Environmental Behavior

Insecticides like dieldrin and endrin have been shown to persist for decades in the environment. A definitive detection of the residues was not possible until 1971 when mass spectrometer started being used as a detector in gas chromatography. Detection of these chemicals in the environment has been reported across the world up to 2005, even though the frequency of reported cases are low due to its relatively small-scale use and very low concentrations.

Endrin regularly enters the environment when applied to crops or when rain washes it off. It has been found in water, sediments, atmospheric air and biotic environment, even after uses have been stopped. Organochlorine pesticides strongly resist degradation, are poorly soluble in water but highly soluble in lipids, which is called lipophilic. This leads to bioaccumulation in fatty tissues of organisms, mainly those dwelling in water. A high bioconcentration factor of 1335-10,000 has been reported in fish. Endrin binds very strongly to organic matter in soil and aquatic sediments due to their high adsorption coefficient, making it less likely to leach into groundwater, even though contaminated groundwater samples have been found. In 2009, EPA released data indicating that the endrin in soil could last up to 14 years or more. The extent of endrin's persistence depends highly on local conditions. For example, high temperature (230 °C) or intense sunlight leads to more rapid breakdown of endrin into endrin ketone and endrin aldehyde, however, this breakdown is less than 5%.

Removal from the Environment

In the United States, endrin was mainly disposed in land until U.S. federal regulations were applied in 1987 on land disposal of wastes containing endrin. Primary methods of endrin disappearance from soil are volatilization and photodecomposition. Under ultraviolet light, endrin forms δ-ketoendrin and International Programme on Chemical Safety (IPCS) claims that in intense summer sun, about 50% of endrin is isomerized to δ-ketoendrin in 7 days. In anaerobic conditions microbial degradation by fungi and bacteria takes place to form the same major end product.

Hazardous Substances Data Bank (HSDB) lists reductive dechlorination and incineration for field disposal of small quantities of endrin. In reductive dechlorination, endrin's chlorine atoms were completely replaced with hydrogen atoms, which is suspected to be more environmentally acceptable. Even though endrin binds very strongly to soil, phytoremediation has been proposed by group of Japanese scientists using crops in the Cucurbitaceae family. As of 2009, exact mechanisms behind the plant uptake of endrin have not been understood. Research in uptake mechanisms and factors that influence the uptake is needed for practical application.

Regulation

United States

In the United States, endrin has been regulated by the EPA. It set a freshwater acute criterion of 0.086 ug/L and a chronic criterion of 0.036 ug/L. In saltwater, the numbers are acute 0.037 and chronic 0.0023 ug/L. The human health contaminate criterion for water plus organism is 0.059 ug/L. The drinking water limit (maximum contaminant level) is set to 2 ppb. Use of endrin in fisheries has been advised against due to the zero tolerance of endrin levels in food products. For occupational exposures to endrin, OSHA and NIOSH have set exposure limits at 0.1 mg/m^3.

International Organizations

The WHO lists Endrin as an obsolete pesticide in its 'Classification of Pesticides by Hazard' and did not assign any hazard class per the Globally Harmonized System of Classification and Labelling of Chemicals.

Taiwan

Taiwan is not a party to the Stockholm Convention as of 2015, but has drafted its own "National Implementation Plan of the Stockholm Convention on Persistent Organic Pollutants" which was approved by the Executive Yuan on April 2008. The Central Competent Authorities of Taiwan sets the limit of 20 mg/kg for soil pollution control. For marine environment quality, standards of 0.002 mg/L has been set. For occupational exposures to endrin, warning has been given that the contact with skin, eyes, and mucous membranes can contribute to the overall exposure.

Heptachlor

Heptachlor is an organochlorine compound that was used as an insecticide. Usually sold as a white or tan powder, heptachlor is one of the cyclodiene insecticides. In 1962, Rachel Carson's *Silent Spring* questioned the safety of heptachlor and other chlorinated insecticides. Due to its highly stable structure, heptachlor can persist in the environment for decades. The US EPA has limited the sale of heptachlor products to the specific application of fire ant control in underground transformers. The amount that can be present in different foods is regulated.

Synthesis

Analogous to the synthesis of other cyclodienes, heptachlor is produced via the Diels-Alder reaction of hexachlorocyclopentadiene and cyclopentadiene. The resulting adduct is brominated followed by treatment with hydrogen chloride in nitromethane in the presence of aluminum trichloride or with iodine monochloride.

Compared to chlordane, it is about 3–5 times more active as an insecticide, but more inert chemically, being resistant to water and caustic alkalies.

Metabolism

Soil microorganisms transform heptachlor by epoxidation, hydrolysis, and reduction. When the compound was incubated with a mixed culture of organisms, chlordene (hexachlorocyclopentadine, its precursor) formed, which was further metabolized to chlordene epoxide. Other metabolites include 1-hydroxychlordene, 1-hydroxy-2,3-epoxychlordene, and heptachlor epoxide. Soil microorganisms hydrolyze heptachlor to give ketochlordene. Rats metabolize heptachlor to the epoxide 1-exo-1-hydroxyheptachlor epoxide and 1,2-dihydrooxydihydrochlordene. When heptachlor epoxide was incubated with microsomal preparations form liver of pigs and from houseflies, the products found were diol and 1-hydroxy-2,3-epoxychlordene. Metabolic scheme in rats shows two pathways with the same metabolite. The first involves following scheme: heptachlor → heptachlor epoxide → dehydrogenated derivative of 1-exo-hydroxy-2,3-exo-epoxychlordene → 1,2-dihydrooxydihydrochlordene. The second involves: Heptachlor → 1-exo-hydroxychlordene → 1-exo-hydroxy, 2,3-exo-epoxychlordene → 1,2-dihydrooxydihydrochlordene.

Environmental Impact

Heptachlor is persistent organic pollutant (POP). It has a half life of ~1.3-4.2 days (air),~0.03-0.11 years (water),~0.11-0.34 years (soil). One study described its half life to be 2 years and claimed that its residues could be found in soil 14 years after its initial application. Like other POPs, heptachlor is lipophilic and poorly soluble in water (0.056 mg/L at 25 °C), thus it tends to accumulate in the body fat of humans and animals.

Heptachlor epoxide is more likely to be found in the environment than its parent compound. The epoxide also dissolves more easily in water than its parent compound and is more persistent. Heptachlor and its epoxide absorb to soil particles and evaporate.

Toxicity of Heptachlor and Related Derivatives

The range of oral rat LD_{50} values are 40 mg/kg to 162 mg/kg. Daily oral doses of heptachlor at 50 and 100 mg/kg were found to be lethal to rats after 10 days. For heptachlor epoxide, the oral LD_{50} values ranging from 46.5 to 60 mg/kg. With rat oral of LD_{50} 47mg/kg, heptachlor epoxide is more toxic. A product of hydrogenation of heptachlor, β-dihydroheptachlor, has high insecticidal activity and low mammalian toxicity, rat oral LD_{50} >5,000mg/kg.

Human Impact

Humans are exposed to heptachlor through drinking water and foods, including breast milk. Heptachlor epoxide is derived from a pesticide that was banned in the U.S. in the 1980s. It is still found in soil and water supplies and can turn up in food and be passed along in breast milk. High levels of it seemed to increase type 2 diabetes risk to about 7 percent (Harmon 2010).

The International Agency for Research on Cancer and the EPA have classified the compound as a possible human carcinogen. Animals exposed to Heptachlor epoxide during gestation and infancy are found to have changes in nervous system and immune function. Higher doses of Heptachlor when exposed to newborn animals caused decrease in body weight and death.

The U.S. EPA MCL for drinking water is 0.0004 mg/L for Heptachlor and 0.0002 mg/L for Heptachlor epoxide. The U.S. FDA limit on food crops is 0.01 ppm, in milk 0.1 ppm, and on edible seafoods 0.3 ppm. The Occupational Safety and Health Administration has limit of 0.5 mg/m^3 (cubic meter of workplace air) for 8-hour shifts and 40-hour work weeks.

An ATSDR report in 1993 found no studies with respect to death in humans after oral exposure to heptachlor or heptachlor epoxide.

Appendix - further Data

Its octanol water coefficient (K_{ow}) is ~$10^{5.27}$. Henry's Law Constant is $2.3 \cdot 10^{-3}$atm-m^3/mol and the vapour pressure is $3 \cdot 10^{-4}$mmHg at 20 °C.

Mirex

Mirex is an organochloride that was commercialized as an insecticide and later banned because of its impact on the environment. This white crystalline odorless solid is a derivative of cyclopentadiene. It was popularized to control fire ants but by virtue of its chemical robustness and lipophilicity it was recognized as a bioaccumulative pollutant. Ironically, the spread of the red imported fire ant was actually encouraged by the use of Mirex, which also kills native ants that are highly competitive with the fire ants. The United States Environmental Protection Agency prohibited its use in 1976.

Production and Applications

Mirex was first synthesized in 1946, but was not used in pesticide formulations until 1955. Mirex was produced by the dimerization of hexachlorocyclopentadiene in the presence of aluminium chloride.

Mirex is a stomach insecticide, meaning that it must be ingested by the organism in order to poison it. The insecticidal use was focused on Southeastern United States to control the imported fire ants *Solenopsis saevissima richteri* and *Solenopsis invicta*. Approximately 250,000 kg of mirex was applied to fields between 1962-75 (US NRC, 1978). Most of the mirex was in the form of "4X mirex bait," which consists of 0.3% mirex in 14.7% soybean oil mixed with 85% corncob grits. Application of the 4X bait was designed to give a coverage of 4.2 g mirex/ha and was delivered by aircraft, helicopter or tractor. 1x and 2x bait were also used. Use of mirex as a pesticide was banned in 1978. The Stockholm Convention banned production and use of several persistent organic pollutant, and Mirex is one of the "dirty dozen".

Degradation

Characteristic of chlorocarbons, mirex does not burn easily; combustion products are expected to include carbon dioxide, carbon monoxide, hydrogen chloride, chlorine, phosgene, and other organochlorine species. Slow oxidation produces chlordecone ("Kepone"), a related insecticide that is also banned in most of the western world, but more readily degraded. Sunlight degrades mirex primarily to photomirex (8-monohydromirex) and later partly to 2,8-dihydromirex.

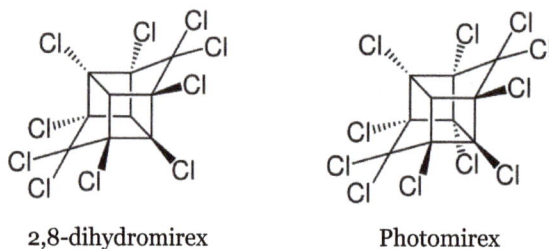

2,8-dihydromirex   Photomirex

Mirex is highly resistant to microbiological degradation. It only slowly dechlorinates to a monohydro derivative by anaerobic microbial action in sewage sludge and by enteric bacteria. Degradation by soil microorganisms has not been described.

Bioaccumulation and Biomagnification

Mirex is highly cumulative and amount depends upon the concentration and duration of exposure. There is evidence of accumulation of mirex in aquatic and terrestrial food chains to harmful levels. After 6 applications of mirex bait at 1.4 kg/ha, high mirex levels were found in some species; turtle fat contained 24.8 mg mirex/kg, kingfishers, 1.9 mg/kg, coyote fat, 6 mg/kg, opossum fat, 9.5 mg/kg, and racoon fat, 73.9 mg/kg. In a model ecosystem with a terrestrial-aquatic interface, sorgum seedlings were treated with mirex at 1.1 kg/ha. Caterpillars fed on these seedlings and their faeces contaminated the water which contained algae, snails, Daphnia, mosquito larvae, and fish. After 33 days, the ecological magnification value was 219 for fish and 1165 for snails.

Although general environmental levels are low, it is widespread in the biotic and abiotic environment. Being lipophilic, Mirex is strongly adsorbed on sediments.

Safety

Mirex is only moderately toxic in single-dose animal studies (oral LD_{50} values range from 365–3000 mg/kg body weight). It can enter the body via inhalation, ingestion, and via the skin. The most sensitive effects of repeated exposure in animals are principally associated with the liver, and these effects have been observed with doses as low as 1.0 mg/kg diet (0.05 mg/kg body weight per day), the lowest dose tested. At higher dose levels, it is fetotoxic (25 mg/kg in diet) and teratogenic (6.0 mg/kg per day). Mirex was not generally active in short-term tests for genetic activity. There is sufficient evidence of its carcinogenicity in mice and rats. Delayed onset of toxic effects and mortality is typical of mirex poisoning. Mirex is toxic for a range of aquatic organisms, with crustacea being particularly sensitive.

Mirex induces pervasive chronic physiological and biochemical disorders in various vertebrates. No acceptable daily intake (ADI) for Mirex has been advised by FAO/WHO. IARC (1979) evaluated mirex's carcinogenic hazard and concluded that "there is sufficient evidence for its carcinogenicity to mice and rats. In the absence of adequate data in humans, based on above result it can be said, that it has carcinogenic risk to humans". Data on human health effects do not exist.

Health Effects

Per a 1995 ATSDR report Mirex caused fatty changes in the livers, hyperexcitability and convulsion, and inhibition of reproduction in animals. It is a potent endocrine disruptor, interfering with estrogen-mediated functions such as ovulation, pregnancy, and endometrial growth. It also induced liver cancer by interaction with estrogen in female rodents.

Toxaphene

Toxaphene was an insecticide used primarily for cotton in the southern United States during the late 1960s and 1970s. Toxaphene is a mixture of over 670 different chemicals and is produced by reacting chlorine gas with camphene. It can be most commonly found as a yellow to amber waxy solid.

Toxaphene was banned in the United States in 1990 and was banned globally by the 2001 Stockholm Convention on Persistent Organic Pollutants. It is a very persistent chemical that can remain in the environment for 1–14 years without degrading, particularly in the soil.

Testing performed on animals, mostly rats and mice, has demonstrated that toxaphene is harmful to animals. Exposure to toxaphene has proven to stimulate the central nervous system, as well as induce morphological changes in the thyroid, liver, and kidneys.

Toxaphene has been shown to cause adverse health effects in humans. The main sources of exposure are through food, drinking water, breathing contaminated air, and direct contact with contaminated soil. Exposure to high levels of toxaphene can cause damage to the lungs, nervous system, liver, kidneys, and in extreme cases, may even cause death. It is thought to be a potential carcinogen in humans, though this has not yet been proven.

Composition

Toxaphene is a synthetic organic compound composed of over 670 chemicals, formed by the chlorination of camphene ($C_{10}H_{16}$) to an overall chlorine content of 67–69% by weight. The bulk of the compounds (mostly chlorobornanes, chlorocamphenes, and other bicyclic chloroorganic compounds) found in toxaphene have chemical formulas ranging from $C_{10}H_{11}Cl_5$ to $C_{10}H_6Cl_{12}$, with a mean formula of $C_{10}H_{10}Cl_8$. The formula weights of these compounds range from 308 to 551 grams/mole; the theoretical mean formula has a value of 414 grams/mole. Toxaphene is usually seen as a yellow to amber waxy solid with a piney odor. It is highly insoluble in water but freely soluble in aromatic hydrocarbons and readily soluble in aliphatic organic solvents. It is stable at room temperature and pressure. It is volatile enough to be transported for long distances through the atmosphere.

Applications

Toxaphene was primarily used as a pesticide for cotton in the southern United States during the late 1960s and 1970s. It was also used on corn, small grains, vegetables, and soybeans to control ectoparasites such as lice, flies, ticks, mange, and scam mites on livestock. In some cases it was used to kill undesirable fish species in lakes and streams. The breakdown of usage can be summarized: 85% on cotton, 7% to control insect pests on livestock and poultry, 5% on other field crops, 3% on soybeans, and less than 1% on sorghum.

The first recorded usage of toxaphene was in 1966 in the United States and by the early to mid 1970's, toxaphene was the United States' most heavily used pesticide. Over 34 million pounds of toxaphene were used annually from 1966 to 1976. As a result of Environmental Protection Agency restrictions, annual toxaphene usage fell to 6.6 million pounds in 1982. In 1990, the EPA banned all usage of toxaphene in the United States. Toxaphene is still used in countries outside the United States but much of this usage has been undocumented. Between 1970 and 1995, global usage of toxaphene was estimated to be 670 million kilograms (1.5 billion pounds).

Production

Toxaphene was first produced in the United States in 1947 although it was not heavily used until 1966. By 1975, toxaphene production reached its peak at 59.4 million pounds annually. Production decreased more than 90% from this value by 1982 due to Environmental Protection Agency restrictions. Overall, an estimated 234,000 metric tons (over 500 million pounds) have been produced in the United States. Between 25% and 35% of the toxaphene produced in the United States has been exported. There are currently 11 toxaphene suppliers worldwide.

Environmental Effects

When released into the environment, toxaphene can be quite persistent and exists in the air, soil, and water. In water, it can evaporate easily and is fairly insoluble. Its solubility is 3 mg/L of water at 22 degrees Celsius. Toxaphene breaks down very slowly and has a half-life of up to 12 years in the soil. It is most commonly found in air, soil, and sediment found at the bottom of lakes or streams. It can also be present in many parts of the world where it was never used because toxaphene is able to evaporate and travel long distances through air currents. Toxaphene can eventually be degraded, through dechlorination, in the air using sunlight to break it down. The degradation of toxaphene usually occurs under aerobic conditions. The levels of toxaphene have decreased since its ban, however, due to its persistence can still be found in the environment today.

Exposure

The three main paths of exposure to toxaphene are ingestion, inhalation, and absorption. For humans, the main source of toxaphene exposure is through ingested seafood. When toxaphene enters the body, it usually accumulates in fatty tissues. It is broken down through dechlorination and oxidation in the liver, and the byproducts are eliminated through feces.

People that live near an area that has high toxaphene contamination are at high risk to toxaphene exposure through inhalation of contaminated air or direct skin contact with contaminated soil or water. Eating large quantities of fish on a daily basis also increases susceptibility to toxaphene exposure. Finally, exposure is rare, yet possible through drinking water when contaminated by toxaphene runoff from the soil. However, toxaphene has been rarely seen at high levels in drinking water due to toxaphene's high levels of insolubility in water.

Shellfish, algae, fish and marine mammals have all been shown to exhibit high levels of toxaphene. People in the Canadian Arctic, where a traditional diet consists of fish and marine animals, have been shown to consume ten times the accepted daily intake of toxaphene. Also, blubber from beluga whales in the Arctic were found to have unhealthy and toxic levels of toxaphene.

Health Effects

Health Effects in Humans

When inhaled or ingested, sufficient quantities of toxaphene can damage the lungs, nervous system, and kidneys, and may cause death. The major health effects of toxaphene involve central nervous system stimulation leading to convulsive seizures. The dose necessary to induce nonfatal convulsions in humans is about 10 milligrams per kilogram body weight per day. Several deaths linked to toxaphene have been documented in which an unknown quantity of toxaphene was ingested intentionally or accidentally from food contamination. The deaths are attributed to respiratory failure resulting from seizures. Chronic inhalation exposure in humans results in reversible respiratory toxicity.

A study conducted between 1954 and 1972 of male agricultural workers and agronomists exposed to toxaphene and other pesticides showed that there are higher proportions of bronchial carcinoma in the test group than in the unexposed general population. However, toxaphene may not have been the main pesticide responsible for tumor production. Tests on lab animals show that toxaphene causes liver and kidney cancer, so the EPA has classified it as a Group B2 carcinogen, meaning it is a probable human carcinogen. The International Agency for Research on Cancer has classified it as a Group 2B carcinogen.

Toxaphene can be detected in blood, urine, breast milk, and body tissues if a person has been exposed to high levels, but it is removed from the body quickly, so detection has to occur within several days of exposure.

It is not known whether toxaphene can affect reproduction in humans.

Health Effects in Animals

Toxaphene was used to treat mange in cattle in California in the 1970s and there were reports of cattle deaths following the toxaphene treatment.

Chronic oral exposure in animals affects the liver, the kidney, the spleen, the adrenal and thyroid glands, the central nervous system, and the immune system. Toxaphene stimulates the central nervous system by antagonizing neurons leading to hyperpolarization of neurons and increased neuronal activity.

Regulations

Toxaphene has been found on at least 68 of the 1,699 National Priorities List sites identified by the United States Environmental Protection Agency. Toxaphene has been forbidden in Germany since 1980. Most uses of toxaphene were cancelled in the U.S. in 1982 with the exception of use on livestock in emergency situations, and for controlling insects on banana and pineapple crops in Puerto Rico and the U.S. Virgin Islands. All uses of toxaphene were cancelled in the U.S. in 1990.

Toxaphene has been banned in 37 countries, including Austria, Belize, Brazil, Costa Rica, Dominican Republic, Egypt, the EU, India, Ireland, Kenya, Korea, Mexico, Panama, Singapore, Thailand and Tonga. Its use has been severely restricted in 11 other countries, including Argentina, Columbia, Dominica, Honduras, Nicaragua, Pakistan, South Africa, Turkey, and Venezuela.

In the Stockholm Convention on POPs, which came into effect on 17 May 2004, twelve POPs were listed to be eliminated or their production and use restricted. The OCPs or pesticide-POPs identified on this list have been termed the "dirty dozen" and include aldrin, chlordane, DDT, dieldrin, endrin, heptachlor, hexachlorobenzene, mirex, and toxaphene.

The EPA has determined that lifetime exposure to 0.01 milligrams per liter of toxaphene in the drinking water is not expected to cause any adverse noncancer effects if the only source of exposure is drinking water, and has established the maximum contaminant level (MCL) of toxaphene at 0.003 mg/L. The United States Food and Drug Administration (FDA) uses the same level for the maximum level permissible in bottled water.

The FDA has determined that the concentration of toxaphene in bottled drinking water should not exceed 0.003 milligrams per liter.

The United States Department of Transportation lists toxaphene as a hazardous material and has special requirements for marking, labeling, and transporting the material.

It is classified as an extremely hazardous substance in the United States as defined in Section 302 of the U.S. Emergency Planning and Community Right-to-Know Act (42 U.S.C. 11002), and is subject to strict reporting requirements by facilities which produce, store, or use it in significant quantities.

Trade Names

Trade names and synonyms include Chlorinated camphene, Octachlorocamphene, Camphochlor, Agricide Maggot Killer, Alltex, Allotox, Crestoxo, Compound 3956, Estonox, Fasco-Terpene, Geniphene, Hercules 3956, M5055, Melipax, Motox, Penphene, Phenacide, Phenatox, Strobane-T, Toxadust, Toxakil, Vertac 90%, Toxon 63, Attac, Anatox, Royal Brand Bean Tox 82, Cotton Tox MP82, Security Tox-Sol-6, Security Tox-MP cotton spray, Security Motox 63 cotton spray, Agro-Chem Brand Torbidan 28, and Dr Roger's TOXENE.

Kepone

Kepone, also known as chlordecone, is an organochlorine compound and a colourless solid. This compound is a controversial insecticide related to Mirex and DDT. Its use was so disastrous that it is now prohibited in the western world, but only after many millions of kilograms had been produced. Kepone is a known persistent organic pollutants (POP), classified among the "dirty dozen" and banned globally by the Stockholm Convention on Persistent Organic Pollutants as of 2011.

Toxicology

The LC50 (LC = lethal concentration) is 0.022–0.095 mg/kg for blue gill and trout. Kepone bioaccumulates in animals by factors up to a million-fold. Workers with repeated exposure suffer severe convulsions resulting from degradation of the synaptic junctions.

Kepone has been found to act as an agonist of the GPER (GPR30).

History

In the US, kepone was produced by Allied Signal Company and LifeSciences Product Company in Hopewell, Virginia. The improper handling and dumping of the substance into the nearby James River (U.S.) in the 1960s and 1970s drew national attention to its toxic effects on humans and wildlife. The product is similar to DDT and is a degradation product of Mirex. The history of Kepone incidents are reviewed in *Who's Poisoning America?: Corporate Polluters and Their Victims in the Chemical Age* (1982). In 2009, Kepone was included in the Stockholm Convention on persistent organic pollutants, which bans its production and use worldwide.

Case Studies

James River Estuary

Due to the pollution risks, many businesses and restaurants along the river suffered economic losses. In 1975 Governor Mills Godwin Jr. shut down the James River to fishing for 100 miles, from Richmond to the Chesapeake Bay. This ban remained in effect for 13 years, until efforts to clean up the river began to get results.

French Antilles

The French island of Martinique is heavily contaminated with kepone, following years of its unrestricted use on banana plantations. Despite a 1990 ban of the substance by France, the economically powerful planter community lobbied intensively to gain the power to continue using kepone until 1993. They had argued that no alternative pesticide was available, which has since been disputed. Similarly, the nearby island of Guadeloupe is also contaminated, but to a lesser extent. Since 2003, local authorities have restricted cultivation of crops because the soil has been seriously contaminated by kepone. Martinique and Guadeloupe have some of the highest prostate cancer diagnosis rates in the world.

In Popular Culture

- Kepone was the name of an American indie rock band from Richmond, Virginia formed in 1991.

- The Dead Kennedys recorded a song named "Kepone Factory", a satire of the controversy surrounding Allied Signal and their negligence regarding employee safety, for their 1981 album *In God We Trust, Inc.*. Written in 1978, the song was originally titled "Kepone Kids".

Synthesis

Kepone is made by dimerizing hexachlorocyclopentadiene and hydrolyzing to a ketone.

Lindane

Lindane, also known as *gamma*-hexachlorocyclohexane, (γ-HCH), gammaxene, Gammallin and sometimes incorrectly called benzene hexachloride (BHC), is an organochlorine chemical variant of hexachlorocyclohexane that has been used both as an agricultural insecticide and as a pharmaceutical treatment for lice and scabies.

Lindane is a neurotoxin that interferes with GABA neurotransmitter function by interacting with the $GABA_A$ receptor-chloride channel complex at the picrotoxin binding site. In humans, lindane affects the nervous system, liver and kidneys, and may well be a carcinogen. It is unclear whether lindane is an endocrine disruptor.

The World Health Organization classifies lindane as "Moderately Hazardous," and its international trade is restricted and regulated under the Rotterdam Convention on Prior Informed Consent. In 2009, the production and agricultural use of lindane was banned under the Stockholm Convention on persistent organic pollutants. A specific exemption to that ban allows it to continue to be used as a second-line pharmaceutical treatment for lice and scabies.

History and Use

The chemical was originally synthesised in 1825 by Faraday. It is named after the Dutch chemist Teunis van der Linden (1884–1965), the first to isolate and describe γ-Hexachlorcyclohexane in 1912. Its pesticidal action was discovered only in 1942, after which lindane production, by Imperial Chemical Industries Ltd (ICI), and use started up in the United Kingdom. It has been used to treat food crops and to forestry products, as a seed treatment, a soil treatment, and to treat livestock and pets. It has also been used as pharmaceutical treatment for lice and scabies, formulated as a shampoo or lotion. It is estimated that between 1950 and 2000, around 600,000 tonnes of lindane were produced globally, and the vast majority of which was used in agriculture. It has been manufactured by several countries, including the United States, China, Brazil, and several European countries, but as of 2007 only India and possibly Russia are still producing it.

By November 2006, the use of lindane had been banned in 52 countries and restricted in 33 others. Seventeen countries, including the US and Canada, allowed either limited agricultural or pharmaceutical use. In 2009, an international ban on the use of lindane in agriculture was implemented under the Stockholm Convention on Persistent Organic Pollutants. A specific exemption allows for it to continue to be used in second-line treatments for the head lice and scabies for five more years. The production of the lindane isomers α- and β-hexachlorocyclohexane was also banned. Although the US has not ratified the Convention, it has similarly banned agricultural uses while still allowing its use a second-line lice and scabies treatment.

United States

In the US, lindane pesticide products were regulated by the U.S. Environmental Protection Agency (EPA), while lindane medications are regulated by the Food and Drug Administration (FDA). It was registered as an agricultural insecticide in the 1940s, and as pharmaceutical in 1951. The EPA gradually began restricting its agricultural use in the 1970s due to concerns over its effects on human health and the environment. By 2002, its use was limited to seed treatments for just 6 crops, and in 2007 these last uses were canceled.

Pharmaceutical Uses

Lindane medications continue to be available in the US, though since 1995 they have been designated "second-line" treatments, meaning they should be prescribed when other "first-line" treatments have failed or cannot be used. In December 2007, the FDA sent a Warning Letter to Morton Grove Pharmaceuticals, the sole U.S. manufacturer of lindane products, requesting that the company correct misleading information on two of its lindane websites. The letter said, in part, that the materials "are misleading in that they omit and/or minimize the most serious and important risk information associated with the use of Lindane Shampoo, particularly in pediatric patients; include a misleading dosing claim; and overstate the efficacy of Lindane Shampoo."

The State of California banned the pharmaceutical lindane, effective 2002, and the Michigan House of Representatives passed a bill in 2009 to restrict its use to doctors' offices. A recent analysis of the California ban concluded that a majority of pediatricians had not experienced problems treating lice or scabies since that ban took effect. The study also documented a marked decrease in lindane wastewater contamination and a dramatic decline in lindane poisoning incidents reported to Poison Control Centers. The authors concluded that, "The California experience suggests elimination of pharmaceutical lindane produced environmental benefits, was associated with a reduction in reported unintentional exposures and did not adversely affect head lice and scabies treatment."

The Persistent Organic Pollutants Review Committee of the Stockholm Convention on Persistent Organic Pollutants considers the use of lindane in agriculture as largely redundant with other, less toxic and less persistent pesticides. In the case of pharmaceutical use, the Committee noted that "alternatives for pharmaceutical uses have often failed for scabies and lice treatment and the number of available alternative products for this use is scarce. For this particular case, a reasonable alternative would be to use lindane as a second-line treatment when other treatments fail, while potential new treatments are assessed."

Human Health Effects

The EPA and WHO both classify lindane as "moderately" acutely toxic. It has an oral LD_{50} of 88 mg/kg in rats and a dermal LD_{50} of 1000 mg/kg. Most of the adverse human health effects reported for lindane have been related to agricultural uses and chronic, occupational exposure of seed-treatment workers.

Exposure to large amounts of lindane can harm the nervous system, producing a range of symptoms from headache and dizziness to seizures, convulsions and, more rarely, death. Lindane has not been shown to affect the immune system in humans and it is not considered to be genotoxic. Prenatal exposure to β-HCH, an isomer of lindane and production byproduct, has been associated with altered thyroid hormone levels and could affect brain development.

The Occupational Safety and Health Administration (OSHA) and National Institute for Occupational Safety and Health (NIOSH) have set occupational exposure limits (permissible exposure and recommended exposure, respectively) for lindane at 0.5 mg/m^3 at a time-weighted average of eight hours for skin exposure. People can be exposed to lindane in the workplace by inhaling it, absorbing it through their skin, swallowing it, and eye contact. At levels of 50 mg/m^3, lindane is immediately dangerous to life and health.

It is classified as an extremely hazardous substance in the United States as defined in Section 302 of the U.S. Emergency Planning and Community Right-to-Know Act (42 U.S.C. 11002), and is subject to strict reporting requirements by facilities which produce, store, or use it in significant quantities.

Cancer Risk

Based primarily on evidence from animal studies, most evaluations of lindane have concluded that it may possibly cause cancer. In 2015, the International Agency for Research on Cancer (IARC) classified lindane as a known human carcinogen, and in 2001 the EPA concluded there was "suggestive evidence of carcinogenicity, but not sufficient to assess human carcinogenic potential." The U.S. Department of Health and Human Services determined that all isomers of hexachlorocyclohexane, including lindane, "may reasonably be anticipated to cause cancer in humans," and in 1999, the EPA characterized the evidence carcinogenicity for lindane as "suggestive ... of carcinogenicity, but not sufficient to assess human carcinogenic potential." Lindane and its isomers have also been on California's Proposition 65 list of known carcinogens since 1989. In contrast, the World Health Organization concluded in 2004 that "lindane is not likely to pose a carcinogenic risk to humans." India's BIS considers Lindane a "confirmed carcinogen".

Adverse Reactions to Lindane

A variety of adverse reactions to lindane pharmaceuticals have been reported, ranging from skin irritation to seizures, and, in rare instances, death. The most common side effects are burning sensations, itching, dryness and rash. While serious effects are rare and have most often resulted from misuse, adverse reactions have occurred when used properly. The FDA therefore requires a so-called black box warning on lindane products, which explains the risks of lindane products and its proper use.

The black box warning emphasizes that lindane should not be used on premature infants and individuals with known uncontrolled seizure disorders, and should be used with caution in infants, children, the elderly, and individuals with other skin conditions (e.g., dermatitis, psoriasis) and people who weigh less than 110 lbs (50 kg) as they may be at risk of serious neurotoxicity.

Environmental Contamination

Lindane is a persistent organic pollutant: it is relatively long-lived in the environment, it is transported long distances by natural processes like global distillation, and it can bioaccumulate in food chains, though it is rapidly eliminated when exposure is discontinued.

The production and agricultural use of lindane are the primary causes of environmental contamination, and levels of lindane in the environment have been decreasing in the U.S., consistent with decreasing agricultural usage patterns. The production of lindane generates large amounts of waste hexachlorocyclohexane isomers, and it is estimated that "every ton of lindane manufactured produces about 9 tons of toxic waste." Modern manufacturing standards for lindane involve the treatment and conversion of waste isomers to less toxic molecules, a process known as "cracking."

When lindane is used in agriculture, an estimated 12–30% of it volatilizes into the atmosphere, where it is subject to long-range transport and can be deposited by rainfall. Lindane in soil can leach to surface and even ground water, and can bioaccumulate in the food chain. However, biotransformation and elimination are relatively rapid when exposure is discontinued. Most exposure of the general population to lindane has resulted from agricultural uses and the intake of foods, such as produce, meats and milk, produced from treated agricultural commodities. Human exposure has decreased significantly since the cancellation of agricultural uses in 2006. Even so, the CDC published in 2005 its Third National Report on Human Exposures to Environmental Chemicals, which found no detectable amounts of lindane in human blood taken from a random sampling of about 5,000 people in the US as part of the NHANES study (National Health and Nutrition Examination Survey at: http://www.cdc.gov/nchs/nhanes/about_nhanes. htm). The lack of detection of lindane in this large human "biomonitoring" study likely reflects the increasingly limited agricultural uses of lindane over the last two decades. The cancellation of agricultural uses in the United States will further reduce the amount of lindane introduced into the environment by more than 99%.

Over time, lindane is broken down in soil, sediment and water into less harmful substances by algae, fungi and bacteria; however, the process is relatively slow and dependent on ambient environmental conditions. The ecological impact of lindane's environmental persistence continues to be debated.

The US EPA determined in 2002 that the Agency does not believe that lindane contaminates drinking water in excess of levels considered safe. U.S. Geological Survey teams concluded the same in 1999 and 2000. With regard to lindane medications, the EPA conducted "down-the-drain" estimates of the amount of lindane reaching public water supplies and concluded that lindane levels from pharmaceutical sources were "extremely low" and not of concern.

Note that the EPA has set the maximum contaminant level or "MCL" for lindane allowed in public water supplies and considered safe for drinking at 200 parts per trillion (ppt). By comparison, the state of California imposes a lower MCL for lindane of 19 ppt. However, the California standard is based on a dated 1988 national water criterion that was subsequently revised by the EPA in 2003 to 980 ppt. The EPA stated that the change resulted from "significant scientific advances made in the last two decades particularly in the areas of cancer and noncancer risk assessments." While the EPA considered raising the MCL standard for lindane to 980 ppt at that time, the change was never implemented because states had little difficultly in maintaining lindane levels below the 200 ppt MCL limit already in place. Today, the legally enforceable MCL standard for lindane is 200 ppt while the national water criterion for lindane is 980 ppt.

Isomers

Lindane is the gamma isomer of hexachlorocyclohexane ("γ-HCH"). In addition to the issue of lindane pollution, there are concerns related to the other isomers of HCH, namely alpha-HCH and beta-HCH, which are notably more toxic than lindane, lack its insecticidal properties, and are byproducts of lindane production. In the 1940s and 1950s, lindane producers stockpiled these isomers in open heaps, which led to ground and water contamination. The International HCH and Pesticide Forum has since been established to bring together experts to address the clean-up and containment of these sites. Modern manufacturing standards for lindane involve the treatment and conversion of waste isomers to less toxic industrial chemicals, a process known as "cracking." Today, only a few production plants remain active worldwide to accommodate public-health uses of lindane and declining agricultural needs. Lindane has not been manufactured in the U.S. since the mid-1970s, but continues to be imported.

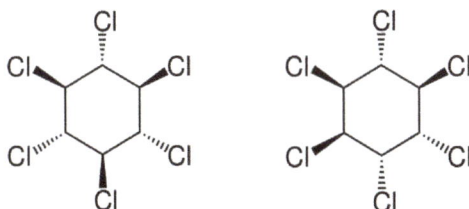

β-hexachlorocyclohexane   α-hexachlorocyclohexane

Endosulfan

Endosulfan is an off-patent organochlorine insecticide and acaricide that is being phased out globally. The two isomers, endo and exo, are known popularly as I and II. Endosulfan sulfate is a product of oxidation containing one extra O atom attached to the S atom. Endosulfan became a highly controversial agrichemical due to its acute toxicity, potential for bioaccumulation, and role as an endocrine disruptor. Because of its threats to human health and the environment, a global ban on the manufacture and use of endosulfan was negotiated under the Stockholm Convention in April 2011. The ban will take effect in mid-2012, with certain uses exempted for five additional years. More than 80 countries, including the European Union, Australia, New Zealand, several West African nations, the United States, Brazil, and Canada had already banned it or announced phase-outs by the time the Stockholm Convention ban was agreed upon. It is still used extensively in India, China, and few other countries. It is produced by Makhteshim Agan and several manufacturers in India and China.

Uses

Endosulfan has been used in agriculture around the world to control insect pests including whiteflies, aphids, leafhoppers, Colorado potato beetles and cabbage worms. Due to its unique mode of action, it is useful in resistance management; however, as it is not specific, it can negatively impact populations of beneficial insects. It is, however, considered to be moderately toxic to honey bees, and it is less toxic to bees than organophosphate insecticides.

Production

The World Health Organization estimated worldwide annual production to be about 9,000 metric tonnes (t) in the early 1980s. From 1980 to 1989, worldwide consumption averaged 10,500 tonnes per year, and for the 1990s use increased to 12,800 tonnes per year.

Endosulfan is a derivative of hexachlorocyclopentadiene, and is chemically similar to aldrin, chlordane, and heptachlor. Specifically, it is produced by the Diels-Alder reaction of hexachlorocyclopentadiene with *cis*-butene-1,4-diol and subsequent reaction of the adduct with thionyl chloride. Technical endosulfan is a 7:3 mixture of stereoisomers, designated α and β. α- and β-Endosulfan are configurational isomers arising from the pyramidal stereochemistry of the teravalent sulfur. α-Endosulfan is the more thermodynamically stable of the two, thus β-endosulfan irreversibly converts to the α form, although the conversion is slow.

History of Commercialization and Regulation

- Early 1950s: Endosulfan was developed.

- 1954: Hoechst AG (now Sanofi) won USDA approval for the use of endosulfan in the United States.

- 2000: Home and garden use in the United States was terminated by agreement with the EPA.

- 2002: The U.S. Fish and Wildlife Service recommended that endosulfan registration should be cancelled, and the EPA determined that endosulfan residues on food and in water pose unacceptable risks. The agency allowed endosulfan to stay on the US market, but imposed restrictions on its agricultural uses.

- 2007: International steps were taken to restrict the use and trade of endosulfan. It is recommended for inclusion in the Rotterdam Convention on Prior Informed Consent, and the European Union proposed inclusion in the list of chemicals banned under the Stockholm Convention on Persistent Organic Pollutants. Such inclusion would ban all use and manufacture of endosulfan globally. Meanwhile, the Canadian government announced that endosulfan was under consideration for phase-out, and Bayer CropScience voluntarily pulled its endosulfan products from the U.S. market but continues to sell the products elsewhere.

- 2008: In February, environmental, consumer, and farm labor groups including the Natural Resources Defense Council, Organic Consumers Association, and the United Farm Workers called on the U.S. EPA to ban endosulfan. In May, coalitions of scientists, environmental groups, and arctic tribes asked the EPA to cancel endosulfan, and in July a coalition of environmental and workers groups filed a lawsuit against the EPA challenging its 2002 decision to not ban it. In October, the Review Committee of the Stockholm Convention moved endosulfan along in the procedure for listing under the treaty, while India blocked its addition to the Rotterdam Convention.

- 2009: The Stockholm Convention's Persistent Organic Pollutants Review Committee (POPRC) agreed that endosulfan is a persistent organic pollutant and that "global action is warranted", setting the stage of a global ban. New Zealand banned endosulfan.

- 2010: The POPRC nominated endosulfan to be added to the Stockholm Convention at the Conference of Parties (COP) in April 2011, which would result in a global ban. The EPA announced that the registration of endosulfan in the U.S. will be cancelled Australia banned the use of the chemical.

- 2011: The Supreme Court of India banned manufacture, sale, and use of toxic pesticide endosulfan in India. The apex court said the ban would remain effective for eight weeks during which an expert committee headed by DG, ICMR, will give an interim report to the court about the harmful effect of the widely used pesticide.

- 2011: the Argentinian Service for Sanity and Agroalimentary Quality (SENASA) decided on August 8 that the import of endosulfan into the South American country will be banned from July 1, 2012 and its commercialization and use from July 1, 2013. In the meantime, a reduced quantity can be imported and sold.

Health Effects

Endosulfan is one of the most toxic pesticides on the market today, responsible for many fatal pesticide poisoning incidents around the world. Endosulfan is also a xenoestrogen—a synthetic substance that imitates or enhances the effect of estrogens—and it can act as an endocrine disruptor, causing reproductive and developmental damage in both animals and humans. It has also been found to act as an aromatase inhibitor. Whether endosulfan can cause cancer is debated. With regard to consumers' intake of endosulfan from residues on food, the Food and Agriculture Organization of United Nations has concluded that long-term exposure from food is unlikely to present a public health concern, but short-term exposure can exceed acute reference doses.

Toxicity

Endosulfan is acutely neurotoxic to both insects and mammals, including humans. The US EPA classifies it as Category I: "Highly Acutely Toxic" based on a LD_{50} value of 30 mg/kg for female rats, while the World Health Organization classifies it as Class II "Moderately Hazardous" based on a rat LD_{50} of 80 mg/kg. It is a GABA-gated chloride channel antagonist, and a Ca^{2+}, Mg^{2+} ATPase inhibitor. Both of these enzymes are involved in the transfer of nerve impulses. Symptoms of acute poisoning include hyperactivity, tremors, convulsions, lack of coordination, staggering, difficulty breathing, nausea and vomiting, diarrhea, and in severe cases, unconsciousness. Doses as low as 35 mg/kg have been documented to cause death in humans, and many cases of sublethal poisoning have resulted in permanent brain damage. Farm workers with chronic endosulfan exposure are at risk of rashes and skin irritation.

EPA's acute reference dose for dietary exposure to endosulfan is 0.015 mg/kg for adults and 0.0015 mg/kg for children. For chronic dietary expsoure, the EPA references doses are 0.006 mg/(kg·day) and 0.0006 mg/(kg·day) for adults and children, respectively.

Endocrine Disruption

Theo Colborn, an expert on endocrine disruption, lists endosulfan as a known endocrine disruptor, and both the EPA and the Agency for Toxic Substances and Disease Registry consider endosulfan to be a potential endocrine disruptor. Numerous *in vitro* studies have documented its potential to disrupt hormones and animal studies have demonstrated its reproductive and developmental toxicity, especially among males. A number of studies have documented that it acts as an antiandrogen in animals. Endosulfan has shown to affect crustacean molt cycles, which are important biological and endocrine-controlled physiological processes essential for the crustacean growth and reproduction. Environmentally relevant doses of endosulfan equal to the EPA's safe dose of 0.006 mg/kg/day have been found to affect gene expression in female rats similarly to the effects of estrogen. It is not known whether endosulfan is a human teratogen (an agent that causes birth defects), though it has significant teratogenic effects in laboratory rats. A 2009 assessment concluded the endocrine disruption in rats occurs only at endosulfan doses that cause neurotoxicity.

Reproductive and Developmental Effects

Several studies have documented that endosulfan can also affect human development. Researchers studying children from many villages in Kasargod District, Kerala, India, have linked endosulfan exposure to delays in sexual maturity among boys. Endosulfan was the only pesticide applied to cashew plantations in the villages for 20 years, and had contaminated the village environment. The researchers compared the villagers to a control group of boys from a demographically similar village that lacked a history of endosulfan pollution. Relative to the control group, the exposed boys had high levels of endosulfan in their bodies, lower levels of testosterone, and delays in reaching sexual maturity. Birth defects of the male reproductive system, including cryptorchidism, were also more prevalent in the study group. The researchers concluded, "our study results suggest that endosulfan exposure in male children may delay sexual maturity and interfere with sex hormone synthesis." Increased incidences of cryptorchidism have been observed in other studies of endosulfan exposed populations.

A 2007 study by the California Department of Public Health found that women who lived near farm fields sprayed with endosulfan and the related organochloride pesticide dicofol during the first eight weeks of pregnancy are several times more likely to give birth to children with autism. This is the first study to look for an association between endosulfan and autism, and additional study is needed to confirm the connection. A 2009 assessment concluded that epidemiology and rodent studies that suggest male reproductive and autism effects are open to other interpretations, and that developmental or reproductive toxicity in rats occurs only at endosulfan doses that cause neurotoxicity.

Endosulfan and Cancer

Endosulfan is not listed as known, probable, or possible carcinogen by the EPA, IARC, or other agencies. No epidemiological studies link exposure to endosulfan specifically to cancer in humans, but *in vitro* assays have shown that endosulfan can promote proliferation of human breast cancer cells. Evidence of carcinogenicity in animals is mixed.

Environmental Fate

Endosulfan is a ubiquitous environmental contaminant. The chemical is semivolatile and persistent to degradation processes in the environment. Endosulfan is subject to

The Government of Gujarat had initiated a study in response to the workers' rally in Bhavnagar and representations made by Sishuvihar, an NGO based in Ahmadabad. The committee constituted for the study also included former Deputy Director of NIOH, Ahmadabad. The committee noted that the WHO, FAO, IARC and US EPA have indicated that endosulfan is not carcinogenic, not teratogenic, not mutagenic and not genotoxic. The highlight of this report is the farmer exposure study based on analysis of their blood reports for residues of endosulfan and the absence of any residues. This corroborates the lack of residues in worker-exposure studies.

The Supreme Court passed interim order on May 13, 2011, in a Writ Petition filed by Democratic Youth Federation of India, (DYFI), a youth wing of Communist Party of India (Marxist) in the backdrop of the incidents reported in Kasargode, Kerala, and banned the production, distribution and use of endosulfan in India because the pesticide has debilitating effects on humans and the environment. The Centre for Science and Environment (CSE) welcomed this order, and called it a 'resounding defeat' for the pesticide industry which has been promoting this deadly toxin. A 2001 study by CSE had established the linkages between the aerial spraying of the pesticide and the growing health disorders in Kasaragode.Over the years, other studies have confirmed these findings, and the health hazards associated with endosulfan are now widely known and accepted. However, in July 2012, the Government asked the Supreme Court to allow use of the pesticide in all states except Kerala and Karnataka, as these states are ready to use it for pest control. India will phase out all endosulfan use by 2017.

New Zealand

Endosulfan was banned in New Zealand by the Environmental Risk Management Authority effective January 2009 after a concerted campaign by environmental groups and the Green Party.

Philippines

A shipment of about 10 tonnes of endosulfan was illegally stowed on the ill-fated MV Princess of the Stars, a ferry that sank off the waters of Romblon (Sibuyan Island), Philippines, during a storm in June 2008. Search, rescue, and salvage efforts were suspended when the endosulfan shipment was discovered, and blood samples from divers at the scene were sent to Malaysia for analysis. The Department of Health of the Philippines has temporarily banned the consumption of fish caught in the area. Endosulfan is classified as a "Severe Marine Pollutant" by the International Maritime Dangerous Goods Code.

United States

In the United States, endosulfan is only registered for agricultural use, and these uses are being phased out. It has been used extensively on cotton, potatoes, tomatoes, and apples

according to the EPA. The EPA estimates that 626 thousand kg of endosulfan were used annually from 1987 to 1997. The US exported more than 140,000 lb of endosulfan from 2001
to 2003, mostly to Latin America, but production and export has since stopped.

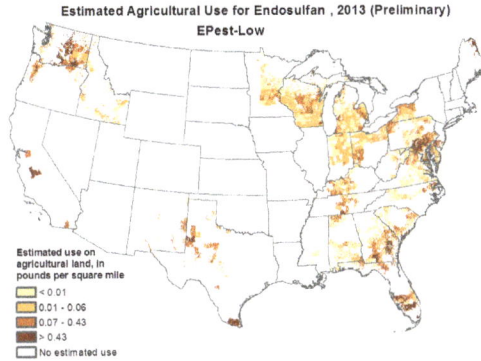

Endosulfan use in the US in pounds per square mile by county in 2013

In California, endosulfan contamination from the San Joaquin Valley has been implicated in the extirpation of the mountain yellow-legged frog from parts of the nearby
Sierra Nevada. In Florida, levels of contamination the Everglades and Biscayne Bay are
high enough to pose a threat to some aquatic organisms.

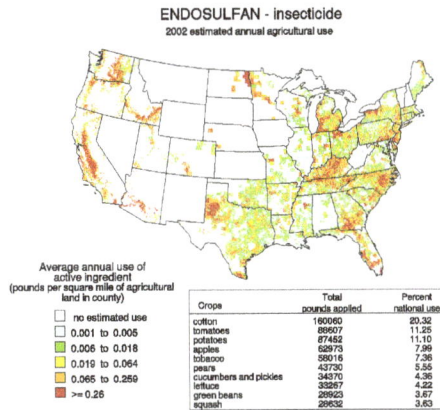

Endosulfan use in the US in pounds per square mile by county in 2002

In 2007, the EPA announced it was rereviewing the safety of endosulfan. The following
year, Pesticide Action Network and NRDC petitioned the EPA to ban endosulfan, and a
coalition of environmental and labor groups sued the EPA seeking to overturn its 2002
decision to not ban endosulfan. In June 2010, the EPA announced it was negotiating a
phaseout of all uses with the sole US manufacturer, Makhteshim Agan, and a complete
ban on the compound.

An official statement by Makhteshim Agan of North America (MANA) states, "From a
scientific standpoint, MANA continues to disagree fundamentally with EPA's conclu-

sions regarding endosulfan and believes that key uses are still eligible for re-registration." The statement adds, "However, given the fact that the endosulfan market is quite small and the cost of developing and submitting additional data high, we have decided to voluntarily negotiate an agreement with EPA that provides growers with an adequate time frame to find alternatives for the damaging insect pests currently controlled by endosulfan."

Australia

Australia banned endosulfan on October 12, 2010, with a two-year phase-out for stock of endosulfan-containing products. Australia had, in 2008, announced endosulfan would not be banned. Citing New Zealand's ban, the Australian Greens called for "zero tolerance" of endosulfan residue on food.

Taiwan

US apples with endosulfan are now allowed to be exported to Taiwan, although the ROC government denied any US pressure on it.

Brazil

Brazil decreed total ban of the substance from July 31, 2013, being forbidden imports of the product from July 31, 2011, date in which national production and utilization begins to be phased out gradually.

DDT

DDT (dichlorodiphenyltrichloroethane) is a colorless, crystalline, tasteless and almost odorless organochlorine known for its insecticidal properties and environmental impacts. DDT has been formulated in multiple forms, including solutions in xylene or petroleum distillates, emulsifiable concentrates, water-wettable powders, granules, aerosols, smoke candles and charges for vaporizers and lotions.

First synthesized in 1874, DDT's insecticidal action was discovered by the Swiss chemist Paul Hermann Müller in 1939. It was used in the second half of World War II to control malaria and typhus among civilians and troops. After the war, DDT was also used as an agricultural insecticide and its production and use duly increased. Müller was awarded the Nobel Prize in Physiology or Medicine "for his discovery of the high efficiency of DDT as a contact poison against several arthropods" in 1948.

In 1962, Rachel Carson's book *Silent Spring* was published. It cataloged the environmental impacts of widespread DDT spraying in the United States and questioned the logic of releasing large amounts of potentially dangerous chemicals into the environment without understanding their effects on the environment or human health. The book claimed that DDT and other pesticides had been shown to cause cancer and that their agricultural use was a threat to wildlife, particularly birds. Its publication was a seminal event for the environmental movement and resulted in a large public outcry that eventually led, in 1972, to a ban on DDT's agricultural use in the United States. A worldwide ban on agricultural use was formalized under the Stockholm Convention on Persistent Organic Pollutants, but its limited and still-controversial use in disease vector control continues, because of its effectiveness in reducing malarial infections, balanced by environmental and other health concerns.

Along with the passage of the Endangered Species Act, the United States ban on DDT is cited by scientists as a major factor in the comeback of the bald eagle (the national bird of the United States) and the peregrine falcon from near-extinction in the contiguous United States.

Properties and Chemistry

DDT is similar in structure to the insecticide methoxychlor and the acaricide dicofol. It is highly hydrophobic and nearly insoluble in water but has good solubility in most organic solvents, fats and oils. DDT does not occur naturally. It is produced by the reaction of chloral (CCl_3CHO) with chlorobenzene (C_6H_5Cl) in the presence of a sulfuric acid catalyst. DDT has been marketed under trade names including Anofex, Cezarex, Chlorophenothane, Clofenotane, Dicophane, Dinocide, Gesarol, Guesapon, Guesarol, Gyron, Ixodex, Neocid, Neocidol and Zerdane.

Isomers and Related Compounds

Commercial DDT is a mixture of several closely–related compounds. The major component (77%) is the p,p' isomer (pictured above). The o,p' isomer (pictured to the right) is also present in significant amounts (15%). Dichlorodiphenyldichloroethylene (DDE) and dichlorodiphenyldichloroethane (DDD) make up the balance. DDE and DDD are the major metabolites and environmental breakdown products. The term "total DDT" is often used to refer to the sum of all DDT related compounds (p,p'-DDT, o,p'-DDT, DDE, and DDD) in a sample.

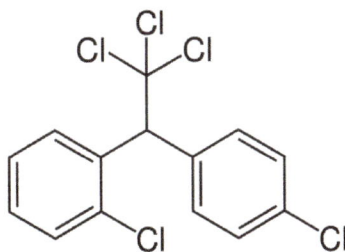

o,p' -DDT, a minor component in commercial DDT.

Production and Use

From 1950 to 1980, DDT was extensively used in agriculture — more than 40,000 tonnes each year worldwide — and it has been estimated that a total of 1.8 million tonnes have been produced globally since the 1940s. In the United States, it was manufactured by some 15 companies, including Monsanto, Ciba, Montrose Chemical Company, Pennwalt and Velsicol Chemical Corporation. Production peaked in 1963 at 82,000 tonnes per year. More than 600,000 tonnes (1.35 billion pounds) were applied in the US before the 1972 ban. Usage peaked in 1959 at about 36,000 tonnes.

In 2009, 3,314 tonnes were produced for malaria control and visceral leishmaniasis. India is the only country still manufacturing DDT and is the largest consumer. China ceased production in 2007.

Mechanism of Insecticide Action

In insects it opens sodium ion channels in neurons, causing them to fire spontaneously, which leads to spasms and eventual death. Insects with certain mutations in their sodium channel gene are resistant to DDT and similar insecticides. DDT resistance is also conferred by up-regulation of genes expressing cytochrome P450 in some insect species, as greater quantities of some enzymes of this group accelerate the toxin's metabolism into inactive metabolites.

History

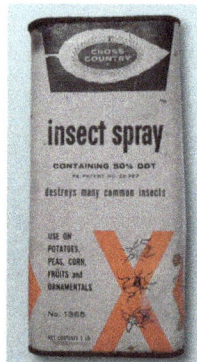

Commercial product concentrate containing 50% DDT, circa 1960s

Commercial product (Powder box, 50 g) containing 10% DDT; Néocide. Ciba Geigy DDT; *"Destroys parasites such as fleas, lice, ants, bedbugs, cockroaches, flies, etc.. Néocide Sprinkle caches of vermin and the places where there are insects and their places of passage. Leave the powder in place as long as possible."* *"Destroy the parasites of man and his dwelling".* *"Death is not instantaneous, it follows inevitably sooner or later."* *"French manufacturing"; "harmless to humans and warm-blooded animals" "sure and lasting effect. Odorless."*

DDT was first synthesized in 1874 by Othmar Zeidler under the supervision of Adolf von Baeyer. It was further described in 1929 in a dissertation by W. Bausch and in two subsequent publications in 1930. The insecticide properties of "multiple chlorinated aliphatic or fat-aromatic alcohols with at least one trichloromethane group" were described in a patent in 1934 by Wolfgang von Leuthold. DDT's insecticidal properties were not, however, discovered until 1939 by the Swiss scientist Paul Hermann Müller, who was awarded the 1948 Nobel Prize in Physiology and Medicine for his efforts.

External audio
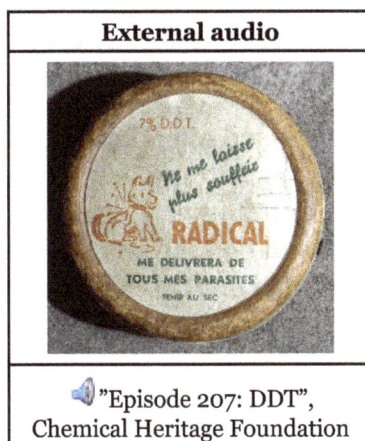
🔊 "Episode 207: DDT", Chemical Heritage Foundation

Use in the 1940s and 1950s

DDT is the best-known of several chlorine-containing pesticides used in the 1940s and 1950s. With pyrethrum in short supply, DDT was used extensively during World War II by the Allies to control the insect vectors of typhus – nearly eliminating the disease in many parts of Europe. In the South Pacific, it was sprayed aerially for malaria and dengue fever control with spectacular effects. While DDT's chemical and insecticidal properties were important factors in these victories, advances in application equipment coupled with competent organization and sufficient manpower were also crucial to the success of these programs.

In 1945, DDT was made available to farmers as an agricultural insecticide and played a role in the final elimination of malaria in Europe and North America.

In 1955, the World Health Organization commenced a program to eradicate malaria in countries with low to moderate transmission rates worldwide, relying largely on DDT

ty is arguably better than the status quo going into the negotiations. For the first time, there is now an insecticide which is restricted to vector control only, meaning that the selection of resistant mosquitoes will be slower than before."

Despite the worldwide ban, agricultural use continued in India, North Korea, and possibly elsewhere as of 2008.

Today, about 3,000 to 4,000 tons of DDT are produced each year for disease vector control. DDT is applied to the inside walls of homes to kill or repel mosquitoes. This intervention, called indoor residual spraying (IRS), greatly reduces environmental damage. It also reduces the incidence of DDT resistance. For comparison, treating 40 hectares (99 acres) of cotton during a typical U.S. growing season requires the same amount of chemical as roughly 1,700 homes.

Environmental Impact

DDT is a persistent organic pollutant that is readily adsorbed to soils and sediments, which can act both as sinks and as long-term sources of exposure affecting organisms. Depending on conditions, its soil half life can range from 22 days to 30 years. Routes of loss and degradation include runoff, volatilization, photolysis and aerobic and anaerobic biodegradation. Due to hydrophobic properties, in aquatic ecosystems DDT and its metabolites are absorbed by aquatic organisms and adsorbed on suspended particles, leaving little DDT dissolved in the water. Its breakdown products and metabolites, DDE and DDD, are also persistent and have similar chemical and physical properties. DDT and its breakdown products are transported from warmer areas to the Arctic by the phenomenon of global distillation, where they then accumulate in the region's food web.

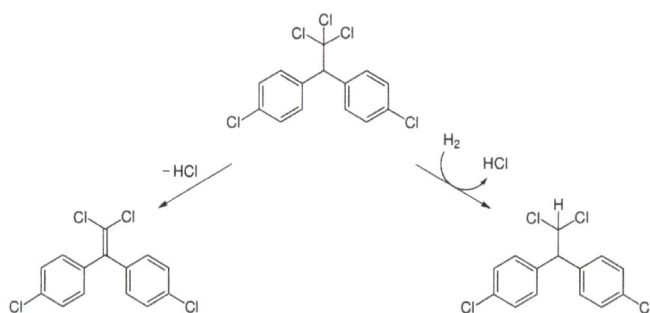

Degradation of DDT to form DDE (by elimination of HCl, left) and DDD (by reductive dechlorination, right)

Because of its lipophilic properties, DDT can bioaccumulate, especially in predatory birds. DDT, DDE and DDD magnify through the food chain, with apex predators such as raptor birds concentrating more chemicals than other animals in the same environment. They are stored mainly in body fat. DDT and DDE are resistant to metabolism; in humans, their half-lives are 6 and up to 10 years, respectively. In the United States, these chemicals were detected in almost all human blood samples tested by the Centers for Disease Control in 2005, though their levels have sharply declined since most uses were banned. Estimated dietary intake has declined, although FDA food tests commonly detect it.

Marine macroalgae (seaweed) help reduce soil toxicity by up to 80% within six weeks.

Effects on Wildlife and Eggshell Thinning

DDT is toxic to a wide range of living organisms, including marine animals such as crayfish, daphnids, sea shrimp and many species of fish. DDE caused eggshell thinning and population declines in multiple North American and European bird of prey species. Eggshell thinning lowers the reproductive success rate of certain bird species by causing egg breakage and embryo deaths. DDE-related eggshell thinning is considered a major reason for the decline of the bald eagle, brown pelican, peregrine falcon and osprey. However, birds vary in their sensitivity to these chemicals. Birds of prey, waterfowl and song birds are more susceptible than chickens and related species. DDE appears to be more potent than DDT. Even in 2010, California condors that feed on sea lions at Big Sur that in turn feed in the Palos Verdes Shelf area of the Montrose Chemical Superfund site exhibited continued thin-shell problems. Scientists with the Ventana Wildlife Society and others study and remediate the condors' problems.

The biological thinning mechanism is not entirely understood, but strong evidence indictates that p,p'-DDE inhibits calcium ATPase in the membrane of the shell gland and reduces the transport of calcium carbonate from blood into the eggshell gland. This results in a dose-dependent thickness reduction. Other evidence indicates that o,p'-DDT disrupts female reproductive tract development, later impairing eggshell quality. Multiple mechanisms may be at work, or different mechanisms may operate in different species. Some studies show that although DDE levels have fallen dramatically, eggshell thickness remains 10–12 percent thinner than before DDT was first used.

Human Health

A U.S. soldier is demonstrating DDT hand-spraying equipment. DDT was used to control the spread of typhus-carrying lice.

DDT is an endocrine disruptor. It is considered likely to be a human carcinogen although the majority of studies suggest it is not directly genotoxic. DDE acts as a weak androgen receptor antagonist, but not as an estrogen. p,p'-DDT, DDT's main component, has little or no androgenic or estrogenic activity. The minor component o,p'-DDT has weak estrogenic activity.

Acute Toxicity

DDT is classified as "moderately toxic" by the US National Toxicology Program (NTP) and "moderately hazardous" by WHO, based on the rat oral LD_{50} of 113 mg/kg. DDT has on rare occasions been administered orally as a treatment for barbiturate poisoning.

Chronic Toxicity

DDT and DDE, like other organochlorines, have been shown to have xenoestrogenic activity, meaning they are chemically similar enough to estrogens to trigger hormonal responses in animals. This endocrine disrupting activity has been observed in mice and rat toxicological studies. Epidemiological evidence indicates that these effects may be occurring in humans as a result of DDT exposure. EPA states that DDT exposure damages the reproductive system and reduces reproductive success. These effects may cause developmental and reproductive toxicity:

- A review article in *The Lancet* states, "research has shown that exposure to DDT at amounts that would be needed in malaria control might cause preterm birth and early weaning ... toxicological evidence shows endocrine-disrupting properties; human data also indicate possible disruption in semen quality, menstruation, gestational length, and duration of lactation."

- Other studies document decreases in semen quality among men with high exposures (generally from IRS).

- Studies generally find that high blood DDT or DDE levels do not increase time to pregnancy (TTP.) Some evidence indicates that the daughters of highly exposed women may have more increased TTP.

- DDT is associated with early pregnancy loss, a type of miscarriage. A prospective cohort study of Chinese textile workers found "a positive, monotonic, exposure-response association between preconception serum total DDT and the risk of subsequent early pregnancy losses." The median serum DDE level of study group was lower than that typically observed in women living in homes sprayed with DDT.

- A Japanese study of congenital hypothyroidism concluded that *in utero* DDT exposure may affect thyroid hormone levels and "play an important role in the incidence and/or causation of cretinism." Other studies found that DDT or DDE interfere with proper thyroid function in pregnancy and childhood.

- Exposure to DDT can cause shorter menstrual cycles.

Carcinogenicity

In 2002, the Centers for Disease Control and Prevention reported, "Overall, in spite of some positive associations for some cancers within certain subgroups of people, there is no clear evidence that exposure to DDT/DDE causes cancer in humans." The NTP classifies it as "reasonably anticipated to be a carcinogen," the International Agency for Research on Cancer classifies it as "probably carcinogenic to humans", and the EPA classifies DDT, DDE and DDD as class B2 "probable" carcinogens. These evaluations are based mainly on animal studies.

A 2005 Lancet review stated that occupational DDT exposure was associated with increased pancreatic cancer risk in 2 case control studies, but another study showed no DDE dose-effect association. Results regarding a possible association with liver cancer and biliary tract cancer are conflicting: workers who did not have direct occupational DDT contact showed increased risk. White men had an increased risk, but not white women or black men. Results about an association with multiple myeloma, prostate and testicular cancer, endometrial cancer and colorectal cancer have been inconclusive or generally do not support an association.

A 2009 review, whose co-authors included persons engaged in DDT-related litigation, reached broadly similar conclusions, with an equivocal association with testicular cancer. Case–control studies did not support an association with leukemia or lymphoma.

Breast Cancer

The question of whether DDT or DDE are risk factors in breast cancer has not been conclusively answered. Several meta analyses of observational studies have concluded that there is no overall relationship between DDT exposure and breast cancer risk. The United States Institute of Medicine reviewed data on the association of breast cancer with DDT exposure in 2012 and concluded that a causative relationship could neither be proven nor disproven.

A 2007 case control study using archived blood samples found that breast cancer risk was increased 5-fold among women who were born prior to 1931 and who had high serum DDT levels in 1963. Reasoning that DDT use became widespread in 1945 and peaked around 1950, they concluded that the ages of 14-20 were a critical period in which DDT exposure leads to increased risk. This study, which suggests a connection between DDT exposure and breast cancer that would not be picked up by most studies, has received variable commentary in third party reviews. One review suggested that "previous studies that measured exposure in older women may have missed the critical period." A second review suggested a cautious approach to the interpretation of these results given methodological weaknesses in the study design. The National Toxicology Program notes that while the majority of studies have not found a relationship between DDT exposure and breast cancer that positive associations have been seen in a "few studies among women with higher levels of exposure and among certain subgroups of women"

A 2015 case control study identified a link (odds ratio 3.4) between *in-utero* exposure (as estimated from archived maternal blood samples) and breast cancer diagnosis in daughters. The findings "support classification of DDT as an endocrine disruptor, a predictor of breast cancer, and a marker of high risk".

Malaria

Malaria remains the primary public health challenge in many countries. 2008 WHO estimates were 243 million cases and 863,000 deaths. About 89% of these deaths occur in Africa, mostly to children under age 5. DDT is one of many tools to fight the disease. Its use in this context has been called everything from a "miracle weapon [that is] like Kryptonite to the mosquitoes," to "toxic colonialism".

Before DDT, eliminating mosquito breeding grounds by drainage or poisoning with Paris green or pyrethrum was sometimes successful. In parts of the world with rising living standards, the elimination of malaria was often a collateral benefit of the introduction of window screens and improved sanitation. A variety of usually simultaneous interventions represents best practice. These include antimalarial drugs to prevent or treat infection; improvements in public health infrastructure to diagnose, sequester and treat infected individuals; bednets and other methods intended to keep mosquitoes from biting humans; and vector control strategies such as larvaciding with insecticides, ecological controls such as draining mosquito breeding grounds or introducing fish to eat larvae and indoor residual spraying (IRS) with insecticides, possibly including DDT. IRS involves the treatment of interior walls and ceilings with insecticides. It is particularly effective against mosquitoes, since many species rest on an indoor wall before or after feeding. DDT is one of 12 WHO–approved IRS insecticides.

WHO's anti-malaria campaign of the 1950s and 1960s relied heavily on DDT and the results were promising, though temporary in developing countries. Experts tie malarial resurgence to multiple factors, including poor leadership, management and funding of malaria control programs; poverty; civil unrest; and increased irrigation. The evolution of resistance to first-generation drugs (e.g. chloroquine) and to insecticides exacerbated the situation. Resistance was largely fueled by unrestricted agricultural use. Resistance and the harm both to humans and the environment led many governments to curtail DDT use in vector control and agriculture. In 2006 WHO reversed a long-standing policy against DDT by recommending that it be used as an indoor pesticide in regions where malaria is a major problem.

Once the mainstay of anti-malaria campaigns, as of 2008 only 12 countries used DDT, including India and some southern African states, though the number was expected to rise.

Initial Effectiveness

When it was introduced in World War II, DDT was effective in reducing malaria morbidity and mortality. WHO's anti-malaria campaign, which consisted mostly of spraying DDT and rapid treatment and diagnosis to break the transmission cycle, was initially successful as well. For example, in Sri Lanka, the program reduced cases from about one million per year before spraying to just 18 in 1963 and 29 in 1964. Thereafter the program was halted to save money and malaria rebounded to 600,000 cases in 1968 and the first quarter of 1969. The country resumed DDT vector control but the mosquitoes had evolved resistance in the interim, presumably because of continued agricultural use. The program switched to malathion, but despite initial successes, malaria continued its resurgence into the 1980s.

DDT remains on WHO's list of insecticides recommended for IRS. After the appointment of Arata Kochi as head of its anti-malaria division, WHO's policy shifted from recommending IRS only in areas of seasonal or episodic transmission of malaria, to advocating it in areas of continuous, intense transmission. WHO reaffirmed its commitment to phasing out DDT, aiming "to achieve a 30% cut in the application of DDT world-wide by 2014 and its total phase-out by the early 2020s if not sooner" while simultaneously combating malaria. WHO plans to implement alternatives to DDT to achieve this goal.

South Africa continues to use DDT under WHO guidelines. In 1996, the country switched to alternative insecticides and malaria incidence increased dramatically. Returning to DDT and introducing new drugs brought malaria back under control. Malaria cases increased in South America after countries in that continent stopped using DDT. Research data showed a strong negative relationship between DDT residual house sprayings and malaria. In a research from 1993 to 1995, Ecuador increased its use of DDT and achieved a 61% reduction in malaria rates, while each of the other countries that gradually decreased its DDT use had large increases.

Mosquito Resistance

In some areas resistance reduced DDT's effectiveness. WHO guidelines require that absence of resistance must be confirmed before using the chemical. Resistance is largely due to agricultural use, in much greater quantities than required for disease prevention.

Resistance was noted early in spray campaigns. Paul Russell, former head of the Allied Anti-Malaria campaign, observed in 1956 that "resistance has appeared after six or seven years." Resistance has been detected in Sri Lanka, Pakistan, Turkey and Central America and it has largely been replaced by organophosphate or carbamate insecticides, *e.g.* malathion or bendiocarb.

In many parts of India, DDT is ineffective. Agricultural uses were banned in 1989 and its anti-malarial use has been declining. Urban use ended. DDT is still manufactured and used. One study concluded that "DDT is still a viable insecticide in indoor residual spraying owing to its effectivity in well supervised spray operation and high excito-repellency factor."

Studies of malaria-vector mosquitoes in KwaZulu-Natal Province, South Africa found susceptibility to 4% DDT (WHO's susceptibility standard), in 63% of the samples, compared to the average of 86.5% in the same species caught in the open. The authors concluded that "Finding DDT resistance in the vector *An. arabiensis*, close to the area where we previously reported pyrethroid-resistance in the vector *An. funestus* Giles, indicates an urgent need to develop a strategy of insecticide resistance management for the malaria control programmes of southern Africa."

DDT can still be effective against resistant mosquitoes and the avoidance of DDT-sprayed walls by mosquitoes is an additional benefit of the chemical. For example, a 2007 study reported that resistant mosquitoes avoided treated huts. The researchers argued that DDT was the best pesticide for use in IRS (even though it did not afford the most protection from mosquitoes out of the three test chemicals) because the others pesticides worked primarily by killing or irritating mosquitoes – encouraging the development of resistance. Others argue that the avoidance behavior slows eradication. Unlike other insecticides such as pyrethroids, DDT requires long exposure to accumulate a lethal dose; however its irritant property shortens contact periods. "For these reasons, when comparisons have been made, better malaria control has generally been achieved with pyrethroids than with DDT." In India outdoor sleeping and night duties are common, implying that "the excito-repellent effect of DDT, often reported useful in other countries, actually promotes outdoor transmission." Genomic studies in the model genetic organism *Drosophila melanogaster* revealed that high level DDT resistance is polygenic, involving multiple resistance mechanisms.

Residents' Concerns

IRS is effective if at least 80% of homes and barns in a residential area are sprayed. Lower coverage rates can jeopardize program effectiveness. Many residents resist DDT spraying, objecting to the lingering smell, stains on walls, and the potential exacerbation of problems with other insect pests. Pyrethroid insecticides (e.g. deltamethrin and lambda-cyhalothrin) can overcome some of these issues, increasing participation.

Human Exposure

A 1994 study found that South Africans living in sprayed homes have levels that are several orders of magnitude greater than others. Breast milk from South African mothers contains high levels of DDT and DDE. It is unclear to what extent these levels arise from home spraying vs food residues. Evidence indicates that these levels are associated with infant neurological abnormalities.

Most studies of DDT's human health effects have been conducted in developed countries where DDT is not used and exposure is relatively low.

Illegal diversion to agriculture is also a concern as it is difficult to prevent and its subsequent use on crops is uncontrolled. For example, DDT use is widespread in Indian agriculture, particularly mango production and is reportedly used by librarians to protect books. Other examples include Ethiopia, where DDT intended for malaria control is reportedly used in coffee production, and Ghana where it is used for fishing." The residues in crops at levels unacceptable for export have been an important factor in bans in several tropical countries. Adding to this problem is a lack of skilled personnel and management.

Criticism of Restrictions on DDT use

Critics argue that limitations on DDT use for public health purposes have caused unnecessary morbidity and mortality from vector-borne diseases, with some claims of malaria deaths ranging as high as the hundreds of thousands and millions. Robert Gwadz of the US National Institutes of Health said in 2007, "The ban on DDT may have killed 20 million children." These arguments were rejected as "outrageous" by former WHO scientist Socrates Litsios. May Berenbaum, University of Illinois entomologist, says, "to blame environmentalists who oppose DDT for more deaths than Hitler is worse than irresponsible." Investigative journalist Adam Sarvana and others characterize this notion as a "myth" promoted principally by Roger Bate of the pro-DDT advocacy group Africa Fighting Malaria (AFM).

Criticisms of a DDT "ban" often specifically reference the 1972 United States ban (with the erroneous implication that this constituted a worldwide ban and prohibited use of DDT in vector control). Reference is often made to *Silent Spring*, even though Carson never pushed for a DDT ban. John Quiggin and Tim Lambert wrote, "the most striking feature of the claim against Carson is the ease with which it can be refuted."

It has been alleged that donor governments and agencies refused to fund DDT spraying, or made aid contingent upon not using DDT. According to a report in the *British Medical Journal*, use of DDT in Mozambique "was stopped several decades ago, because 80% of the country's health budget came from donor funds, and donors refused to allow the use of DDT." Roger Bate asserted, "many countries have been coming under pressure from international health and environment agencies to give up DDT or face losing aid grants: Belize and Bolivia are on record admitting they gave in to pressure on this issue from [USAID]."

The US Agency for International Development (USAID) has been the focus of much criticism. While the agency now funds DDT use in some African countries, in the past it did not. When John Stossel accused USAID of not funding DDT because it wasn't "politically correct," Anne Peterson, the agency's assistant administrator for global health, replied that "I believe that the strategies we are using are as effective as spraying with DDT ... So, politically correct or not, I am very confident that what we are doing is the right strategy." USAID's Kent R. Hill stated that the agency had been misrepresented: "USAID strongly supports spraying as a preventative measure for malaria and will support the use of DDT when it is scientifically sound and warranted." The Agency's website states that "USAID has never had a 'policy' as such either 'for' or 'against' DDT for IRS (Indoor residual spraying). The real change in the past two years [2006/07] was a new interest and emphasis on IRS in general – with DDT or any other insecticide – as an effective malaria prevention strategy in tropical Africa." The agency claimed that in many cases alternative malaria control measures were more cost-effective than DDT spraying.

Alternatives

Insecticides

Organophosphate and carbamate insecticides, *e.g.* malathion and bendiocarb, respectively, are more expensive than DDT per kilogram and are applied at roughly the same dosage. Pyrethroids such as deltamethrin are also more expensive than DDT, but are applied more sparingly ($0.02–0.3 \text{ g/m}^2$ vs $1–2 \text{ g/m}^2$), so the net cost per house is about the same.

Non-chemical Vector Control

Before DDT, malaria was successfully eliminated or curtailed in several tropical areas by removing or poisoning mosquito breeding grounds and larva habitats, for example by eliminating standing water. These methods have seen little application in Africa for more than half a century. According to CDC, such methods are not practical in Africa because "*Anopheles gambiae*, one of the primary vectors of malaria in Africa, breeds in numerous small pools of water that form due to rainfall ... It is difficult, if not impossible, to predict when and where the breeding sites will form, and to find and treat them before the adults emerge."

The relative effectiveness of IRS versus other malaria control techniques (e.g. bednets or prompt access to anti-malarial drugs) varies and is dependent on local conditions.

A WHO study released in January 2008 found that mass distribution of insecticide-treated mosquito nets and artemisinin–based drugs cut malaria deaths in half in malaria-burdened Rwanda and Ethiopia. IRS with DDT did not play an important role in mortality reduction in these countries.

Vietnam has enjoyed declining malaria cases and a 97% mortality reduction after switching in 1991 from a poorly funded DDT-based campaign to a program based on prompt treatment, bednets and pyrethroid group insecticides.

In Mexico, effective and affordable chemical and non-chemical strategies were so successful that the Mexican DDT manufacturing plant ceased production due to lack of demand.

A review of fourteen studies in sub-Saharan Africa, covering insecticide-treated nets, residual spraying, chemoprophylaxis for children, chemoprophylaxis or intermittent treatment for pregnant women, a hypothetical vaccine and changing front–line drug treatment, found decision making limited by the lack of information on the costs and effects of many interventions, the small number of cost-effectiveness analyses, the lack of evidence on the costs and effects of packages of measures and the problems in generalizing or comparing studies that relate to specific settings and use different methodologies and outcome measures. The two cost-effectiveness estimates of DDT residual spraying examined were not found to provide an accurate estimate of the cost-effectiveness of DDT spraying; the resulting estimates may not be good predictors of cost-effectiveness in current programs.

However, a study in Thailand found the cost per malaria case prevented of DDT spraying (US$1.87) to be 21% greater than the cost per case prevented of lambda-cyhalothrin–treated nets (US$1.54), casting some doubt on the assumption that DDT was the most cost-effective measure. The director of Mexico's malaria control program found similar results, declaring that it was 25% cheaper for Mexico to spray a house with synthetic pyrethroids than with DDT. However, another study in South Africa found generally lower costs for DDT spraying than for impregnated nets.

A more comprehensive approach to measuring cost-effectiveness or efficacy of malarial control would not only measure the cost in dollars, as well as the number of people saved, but would also consider ecological damage and negative human health impacts. One preliminary study found that it is likely that the detriment to human health approaches or exceeds the beneficial reductions in malarial cases, except perhaps in epidemics. It is similar to the earlier study regarding estimated theoretical infant mortality caused by DDT and subject to the criticism also mentioned earlier.

A study in the Solomon Islands found that "although impregnated bed nets cannot entirely replace DDT spraying without substantial increase in incidence, their use permits reduced DDT spraying."

A comparison of four successful programs against malaria in Brazil, India, Eritrea and Vietnam does not endorse any single strategy but instead states, "Common success factors included conducive country conditions, a targeted technical approach using a package of effective tools, data-driven decision-making, active leadership at all levels of government, involvement of communities, decentralized implementation and control of finances, skilled technical and managerial capacity at national and sub-national levels, hands-on technical and programmatic support from partner agencies, and sufficient and flexible financing."

DDT resistant mosquitoes have generally proved susceptible to pyrethroids. Thus far, pyrethroid resistance in *Anopheles* has not been a major problem.

Fungicide

Fungicides are biocidal chemical compounds or biological organisms used to kill fungi or fungal spores. A fungistatic inhibits their growth. Fungi can cause serious damage in agriculture, resulting in critical losses of yield, quality, and profit. Fungicides are used both in agriculture and to fight fungal infections in animals. Chemicals used to control oomycetes, which are not fungi, are also referred to as fungicides, as oomycetes use the same mechanisms as fungi to infect plants.

Fungicides can either be contact, translaminar or systemic. Contact fungicides are not taken up into the plant tissue and protect only the plant where the spray is deposited. Translaminar fungicides redistribute the fungicide from the upper, sprayed leaf surface to the lower, unsprayed surface. Systemic fungicides are taken up and redistributed through the xylem vessels. Few fungicides move to all parts of a plant. Some are locally systemic, and some move upwardly.

Most fungicides that can be bought retail are sold in a liquid form. A very common active ingredient is sulfur, present at 0.08% in weaker concentrates, and as high as 0.5% for more potent fungicides. Fungicides in powdered form are usually around 90% sulfur and are very toxic. Other active ingredients in fungicides include neem oil, rosemary oil, jojoba oil, the bacterium *Bacillus subtilis*, and the beneficial fungus *Ulocladium oudemansii*.

Fungicide residues have been found on food for human consumption, mostly from post-harvest treatments. Some fungicides are dangerous to human health, such as vinclozolin, which has now been removed from use. Ziram is also a fungicide that is thought to be toxic to humans if exposed to chronically. A number of fungicides are also used in human health care.

Natural Fungicides

Plants and other organisms have chemical defenses that give them an advantage against microorganisms such as fungi. Some of these compounds can be used as fungicides:

- Tea tree oil

- Cinnamaldehyde

- Citronella oil

- Jojoba oil

- Nimbin

- Oregano oil

- Rosemary oil

- Monocerin

- Milk

Whole live or dead organisms that are efficient at killing or inhibiting fungi can sometimes be used as fungicides:

- *Bacillus subtilis*

- *Ulocladium oudemansii*

- Kelp (powdered dried kelp is fed to cattle to help prevent fungal infection)

- *Ampelomyces quisqualis*

Resistance

Pathogens respond to the use of fungicides by evolving resistance. In the field several mechanisms of resistance have been identified. The evolution of fungicide resistance can be gradual or sudden. In qualitative or discrete resistance, a mutation (normally to a single gene) produces a race of a fungus with a high degree of resistance. Such resistant varieties also tend to show stability, persisting after the fungicide has been removed from the market. For example, sugar beet leaf blotch remains resistant to azoles years after they were no longer used for control of the disease. This is because such mutations often have a high selection pressure when the fungicide is used, but there is low selection pressure to remove them in the absence of the fungicide.

In instances where resistance occurs more gradually, a shift in sensitivity in the pathogen to the fungicide can be seen. Such resistance is polygenic – an accumulation of many mutations in different genes, each having a small additive effect. This type of resistance is known as quantitative or continuous resistance. In this kind of resistance, the pathogen population will revert to a sensitive state if the fungicide is no longer applied.

Little is known about how variations in fungicide treatment affect the selection pressure to evolve resistance to that fungicide. Evidence shows that the doses that provide the most control of the disease also provide the largest selection pressure to acquire resistance, and that lower doses decrease the selection pressure.

In some cases when a pathogen evolves resistance to one fungicide, it automatically obtains resistance to others – a phenomenon known as cross resistance. These additional fungicides are normally of the same chemical family or have the same mode of action, or can be detoxified by the same mechanism. Sometimes negative cross resistance occurs, where resistance to one chemical class of fungicides leads to an increase in sensitivity to a different chemical class of fungicides. This has been seen with carbendazim and diethofencarb.

There are also recorded incidences of the evolution of multiple drug resistance by pathogens – resistance to two chemically different fungicides by separate mutation events. For example, *Botrytis cinerea* is resistant to both azoles and dicarboximide fungicides.

There are several routes by which pathogens can evolve fungicide resistance. The most common mechanism appears to be alteration of the target site, in particular as a defence against single site of action fungicides. For example, Black Sigatoka, an economically important pathogen of banana, is resistant to the QoI fungicides, due to a single nucleotide change resulting in the replacement of one amino acid (glycine) by another (alanine) in the target protein of the QoI fungicides, cytochrome b. It is presumed that this disrupts the binding of the fungicide to the protein, rendering the fungicide ineffective. Upregulation of target genes can also render the fungicide ineffective. This is seen in DMI-resistant strains of *Venturia inaequalis*.

Resistance to fungicides can also be developed by efficient efflux of the fungicide out of the cell. *Septoria tritici* has developed multiple drug resistance using this mechanism. The pathogen had 5 ABC-type transporters with overlapping substrate specificities that together work to pump toxic chemicals out of the cell.

In addition to the mechanisms outlined above, fungi may also develop metabolic pathways that circumvent the target protein, or acquire enzymes that enable metabolism of the fungicide to a harmless substance.

Fungicide Resistance Management

The fungicide resistance action committee (FRAC) has several recommended practices to try to avoid the development of fungicide resistance, especially in at-risk fungicides including *Strobilurins* such as azoxystrobin.

Products should not be used in isolation, but rather as mixture, or alternate sprays, with another fungicide with a different mechanism of action. The likelihood of the pathogen's developing resistance is greatly decreased by the fact that any resistant isolates to one fungicide will be killed by the other; in other words, two mutations would be required rather than just one. The effectiveness of this technique can be demonstrated by Metalaxyl, a phenylamide fungicide. When used as the sole product in Ireland to control potato blight (*Phytophthora infestans*), resistance developed within one growing season. However, in countries like the UK where it was marketed only as a mixture, resistance problems developed more slowly.

Fungicides should be applied only when absolutely necessary, especially if they are in an at-risk group. Lowering the amount of fungicide in the environment lowers the selection pressure for resistance to develop.

Manufacturers' doses should always be followed. These doses are normally designed to give the right balance between controlling the disease and limiting the risk of resistance development. Higher doses increase the selection pressure for single-site mutations that confer resistance, as all strains but those that carry the mutation will be eliminated, and thus the resistant strain will propagate. Lower doses greatly increase the risk of polygenic resistance, as strains that are slightly less sensitive to the fungicide may survive.

It is also recommended that where possible fungicides are used only in a protective manner, rather than to try to cure already-infected crops. Far fewer fungicides have curative/eradicative ability than protectant. Thus, fungicide preparations advertised as having curative action may have only one active chemical; a single fungicide acting in isolation increases the risk of fungicide resistance.

It is better to use an integrative pest management approach to disease control rather than relying on fungicides alone. This involves the use of resistant varieties and hygienic practices, such as the removal of potato discard piles and stubble on which the pathogen can overwinter, greatly reducing the titre of the pathogen and thus the risk of fungicide resistance development.

Hexachlorobenzene

Hexachlorobenzene, or perchlorobenzene, is an organochloride with the molecular formula C_6Cl_6. It is a fungicide formerly used as a seed treatment, especially on wheat to control the fungal disease bunt. It has been banned globally under the Stockholm Convention on persistent organic pollutants.

Physical and Chemical Properties

HCB is a white crystalline solid that has negligible solubility in water (0.00000002 M) and variable solubility in different organic solvents. It is most soluble in halogenated solvents like chloroform (approx 0.03 M) less soluble in esters and hydrocarbons (approx 0.020 M), and even less soluble in short chain alcohols (0.002-0.006 M). Its vapour pressure is 1.09×10^{-5} mmHg (1.45 mPa) at 20 °C. Its flash point is 242 °C and it sublimes at 322 °C.

Safety

Hexachlorobenzene is an animal carcinogen and is considered to be a probable human carcinogen. After its introduction as a fungicide in 1945, for crop seeds, this toxic chemical was found in all food types. Hexachlorobenzene was banned from use in the United States in 1966.

This material has been classified by the International Agency for Research on Cancer (IARC) as a Group 2B carcinogen (possibly carcinogenic to humans). Animal carcinogenicity data for hexachlorobenzene show increased incidences of liver, kidney (renal tubular tumours) and thyroid cancers. Chronic oral exposure in humans has been shown to give rise to a liver disease (porphyria cutanea tarda), skin lesions with discoloration, ulceration, photosensitivity, thyroid effects, bone effects and loss of hair. Neurological changes have been reported in rodents exposed to hexachlorobenzene. Hexachlorobenzene may cause embryolethality and teratogenic effects. Human and animal studies have demonstrated that hexachlorobenzene crosses the placenta to accumulate in foetal tissues and is transferred in breast milk.

HCB is very toxic to aquatic organisms. It may cause long term adverse effects in the aquatic environment. Therefore, release into waterways should be avoided. It is persistent in the environment. Ecological investigations have found that biomagnification up the food chain does occur. Hexachlorobenzene has a half life in the soil of between 3 and 6 years. Risk of bioaccumulation in an aquatic species is high.

Toxicology

- Oral LD50 (rat): 10,000 mg/kg

- Oral LD50 (mice): 4,000 mg/kg

- Inhalation LC50 (rat): 3600 mg/m^3

Material has relatively low acute toxicity but is toxic because of its persistent and cumulative nature in body tissues in rich lipid content.

Unique Exposure Incident

In Anatolia, Turkey between 1955 and 1959, during a period when bread wheat was unavailable, 500 people were fatally poisoned and more than 4,000 people fell ill by eating bread made with HCB-treated seed that was intended for agriculture use. Most of the sick were affected with a liver condition called porphyria cutanea tarda, which disturbs the metabolism of hemoglobin and results in skin lesions. Almost all breast-feeding children under the age of two, whose mothers had eaten tainted bread, died from a condition called "pembe yara" or "pink sore," most likely from high doses of HCB in the breast milk. In one mother's breast milk the HCB level was found to be 20 parts per million in lipid, approximately 2,000 times the average levels of contamination found in breast-milk samples around the world. Follow-up studies 20 to 30 years after the poisoning found average HCB levels in breast milk were still more than seven times the average for unexposed women in that part of the world (56 specimens of human milk obtained from mothers with porphyria, average value was 0.51 ppm in HCB-exposed patients compared to 0.07 ppm in unexposed controls), and 150 times the level allowed in cow's milk.

In the same follow-up study of 252 patients (162 males and 90 females, avg. current age of 35.7 years), 20–30 years postexposure, many subjects had dermatologic, neurologic, and orthopedic symptoms and signs. The observed clinical findings include scarring of the face and hands (83.7%), hyperpigmentation (65%), hypertrichosis (44.8%), pinched faces (40.1%), painless arthritis (70.2%), small hands (66.6%), sensory shading (60.6%), myotonia (37.9%), cogwheeling (41.9%), enlarged thyroid (34.9%), and enlarged liver (4.8%). Urine and stool porphyrin levels were determined in all patients, and 17 have at least one of the porphyrins elevated. Offspring of mothers with three decades of HCB-induced porphyria appear normal.

Pesticide

Pesticides are substances meant for attracting, seducing, and then destroying any pest. They are a class of biocide. The most common use of pesticides is as plant protection products (also known as crop protection products), which in general protect plants from damaging influences such as weeds, fungi, or insects. This use of pesticides is so common that the term *pesticide* is often treated as synonymous with *plant protection product*, although it is in fact a broader term, as pesticides are also used for non-agricultural purposes. The term pesticide includes all of the following: herbicide, insecticide, insect growth regulator, nematicide, termiticide, molluscicide, piscicide, avicide, rodenticide, predacide, bactericide, insect repellent, animal repellent, antimicrobial, fungicide, disinfectant (antimicrobial), and sanitizer.

A crop-duster spraying pesticide on a field

In general, a pesticide is a chemical or biological agent (such as a virus, bacterium, antimicrobial, or disinfectant) that deters, incapacitates, kills, or otherwise discourages pests. Target pests can include insects, plant pathogens, weeds, mollusks, birds, mammals, fish, nematodes (roundworms), and microbes that destroy property, cause nuisance, or spread disease, or are disease vectors. Although pesticides have benefits, some also have drawbacks, such as potential toxicity to humans and other species. According to the Stockholm Convention on Persistent Organic Pollutants, 9 of the 12 most dangerous and persistent organic chemicals are organochlorine pesticides.

A Lite-Trac four-wheeled self-propelled crop sprayer spraying pesticide on a field

Definition

The Food and Agriculture Organization (FAO) has defined *pesticide* as:

Type of pesticide	Target pest group
Herbicides	Plant
Algicides or Algaecides	Algae
Avicides	Birds
Bactericides	Bacteria
Fungicides	Fungi and Oomycetes
Insecticides	Insects
Miticides or Acaricides	Mites
Molluscicides	Snails
Nematicides	Nematodes
Rodenticides	Rodents
Virucides	Viruses

any substance or mixture of substances intended for preventing, destroying, or controlling any pest, including vectors of human or animal disease, unwanted species of plants or animals, causing harm during or otherwise interfering with the production, processing, storage, transport, or marketing of food, agricultural commodities, wood and wood products or animal feedstuffs, or substances that may be administered to animals for the control of insects, arachnids, or other pests in or on their bodies. The term includes substances intended for use as a plant growth regulator, defoliant, desiccant, or agent for thinning fruit or preventing the premature fall of fruit. Also used as substances applied to crops either before or after harvest to protect the commodity from deterioration during storage and transport.

Pesticides can be classified by target organism (e.g., herbicides, insecticides, fungicides, rodenticides, and pediculicides - see table), chemical structure (e.g., organic, inorganic, synthetic, or biological (biopesticide), although the distinction can sometimes blur), and physical state (e.g. gaseous (fumigant)). Biopesticides include microbial pesticides and biochemical pesticides. Plant-derived pesticides, or "botanicals", have been developing quickly. These include the pyrethroids, rotenoids, nicotinoids, and a fourth group that includes strychnine and scilliroside.

Many pesticides can be grouped into chemical families. Prominent insecticide families include organochlorines, organophosphates, and carbamates. Organochlorine hydrocarbons (e.g., DDT) could be separated into dichlorodiphenylethanes, cyclodiene compounds, and other related compounds. They operate by disrupting the sodium/potassium balance of the nerve fiber, forcing the nerve to transmit continuously. Their toxicities vary greatly, but they have been phased out because of their persistence and potential to bioaccumulate. Organophosphate and carbamates largely replaced organochlorines. Both operate through inhibiting the enzyme acetylcholinesterase, allowing acetylcholine to transfer nerve impulses indefinitely and causing a variety of symptoms such as weakness or paralysis. Organophosphates are quite toxic to vertebrates, and have in some cases been replaced by less toxic carbamates. Thiocarbamate and dithiocarbamates are subclasses of carbamates. Prominent families of herbicides include phenoxy and benzoic acid herbicides (e.g. 2,4-D), triazines (e.g., atrazine), ureas (e.g., diuron), and Chloroacetanilides (e.g., alachlor). Phenoxy compounds tend to selectively kill broad-leaf weeds rather than grasses. The phenoxy and benzoic acid herbicides function similar to plant growth hormones, and grow cells without normal cell division, crushing the plant's nutrient transport system. Triazines interfere with photosynthesis. Many commonly used pesticides are not included in these families, including glyphosate.

Pesticides can be classified based upon their biological mechanism function or application method. Most pesticides work by poisoning pests. A systemic pesticide moves inside a plant following absorption by the plant. With insecticides and most fungicides, this movement is usually upward (through the xylem) and outward. Increased efficiency may be a result. Systemic insecticides, which poison pollen and nectar in the flowers, may kill bees and other needed pollinators.

In 2009, the development of a new class of fungicides called paldoxins was announced. These work by taking advantage of natural defense chemicals released by plants called phytoalexins, which fungi then detoxify using enzymes. The paldoxins inhibit the fungi's detoxification enzymes. They are believed to be safer and greener.

Uses

Pesticides are used to control organisms that are considered to be harmful. For example, they are used to kill mosquitoes that can transmit potentially deadly diseases like West Nile virus, yellow fever, and malaria. They can also kill bees, wasps or ants that can cause allergic reactions. Insecticides can protect animals from illnesses that can be caused by parasites such as fleas. Pesticides can prevent sickness in humans that could be caused by moldy food or diseased produce. Herbicides can be used to clear roadside weeds, trees and brush. They can also kill invasive weeds that may cause environmental damage. Herbicides are commonly applied in ponds and lakes to control algae and plants such as water grasses that can interfere with activities like swimming and fishing and cause the water to look or smell unpleasant. Uncontrolled pests such as termites and mold can damage structures such as houses. Pesticides are used in grocery stores and food storage facilities to manage rodents and insects that infest food such as grain. Each use of a pesticide carries some associated risk. Proper pesticide use decreases these associated risks to a level deemed acceptable by pesticide regulatory agencies such as the United States Environmental Protection Agency (EPA) and the Pest Management Regulatory Agency (PMRA) of Canada.

DDT, sprayed on the walls of houses, is an organochlorine that has been used to fight malaria since the 1950s. Recent policy statements by the World Health Organization have given stronger support to this approach. However, DDT and other organochlorine pesticides have been banned in most countries worldwide because of their persistence in the environment and human toxicity. DDT use is not always effective, as resistance to DDT was identified in Africa as early as 1955, and by 1972 nineteen species of mosquito worldwide were resistant to DDT.

Amount Used

In 2006 and 2007, the world used approximately 2.4 megatonnes (5.3×10^9 lb) of pesticides, with herbicides constituting the biggest part of the world pesticide use at 40%, followed by insecticides (17%) and fungicides (10%). In 2006 and 2007 the U.S. used approximately 0.5 megatonnes (1.1×10^9 lb) of pesticides, accounting for 22% of the world total, including 857 million pounds (389 kt) of conventional pesticides, which are used in the agricultural sector (80% of conventional pesticide use) as well as the industrial, commercial, governmental and home & garden sectors.Pesticides are also found in majority of U.S. households with 78 million out of the 105.5 million households indicating that they use some form of pesticide. As of 2007, there were more than 1,055 active ingredients registered as pesticides, which yield over 20,000 pesticide products that are marketed in the United States.

The US used some 1 kg (2.2 pounds) per hectare of arable land compared with: 4.7 kg in China, 1.3 kg in the UK, 0.1 kg in Cameroon, 5.9 kg in Japan and 2.5 kg in Italy. Insecticide use in the US has declined by more than half since 1980, (.6%/yr) mostly due to the near phase-out of organophosphates. In corn fields, the decline was even steeper, due to the switchover to transgenic Bt corn.

For the global market of crop protection products, market analysts forecast revenues of over 52 billion US$ in 2019.

Benefits

Pesticides can save farmers' money by preventing crop losses to insects and other pests; in the U.S., farmers get an estimated fourfold return on money they spend on pesticides. One study found that not using pesticides reduced crop yields by about 10%. Another study, conducted in 1999, found that a ban on pesticides in the United States may result in a rise of food prices, loss of jobs, and an increase in world hunger.

There are two levels of benefits for pesticide use, primary and secondary. Primary benefits are direct gains from the use of pesticides and secondary benefits are effects that are more long-term.

Primary Benefits

1. Controlling pests and plant disease vectors

    o Improved crop/livestock yields

    o Improved crop/livestock quality

    o Invasive species controlled

2. Controlling human/livestock disease vectors and nuisance organisms

    o Human lives saved and suffering reduced

    o Animal lives saved and suffering reduced

    o Diseases contained geographically

3. Controlling organisms that harm other human activities and structures

    o Drivers view unobstructed

    o Tree/brush/leaf hazards prevented

    o Wooden structures protected

Monetary

Every dollar ($1) that is spent on pesticides for crops yields four dollars ($4) in crops saved. This means based that, on the amount of money spent per year on pesticides, $10 billion, there is an additional $40 billion savings in crop that would be lost due to damage by insects and weeds. In general, farmers benefit from having an increase in crop yield and from being able to grow a variety of crops throughout the year. Consumers of agricultural products also benefit from being able to afford the vast quantities of produce available year-round. The general public also benefits from the use of pesticides for the control of insect-borne diseases and illnesses, such as malaria. The use of pesticides creates a large job market within the agrichemical sector.

Costs

On the cost side of pesticide use there can be costs to the environment, costs to human health, as well as costs of the development and research of new pesticides.

Health Effects

Pesticides may cause acute and delayed health effects in people who are exposed. Pesticide exposure can cause a variety of adverse health effects, ranging from simple irritation of the skin and eyes to more severe effects such as affecting the nervous system, mimicking hormones causing reproductive problems, and also causing cancer. A 2007 systematic review found that "most studies on non-Hodgkin lymphoma and leukemia showed positive associations with pesticide exposure" and thus concluded that cosmetic use of pesticides should be decreased. There is substantial evidence of associations between organophosphate insecticide exposures and neurobehavioral alterations. Limited evidence also exists for other negative outcomes from pesticide exposure including neurological, birth defects, and fetal death.

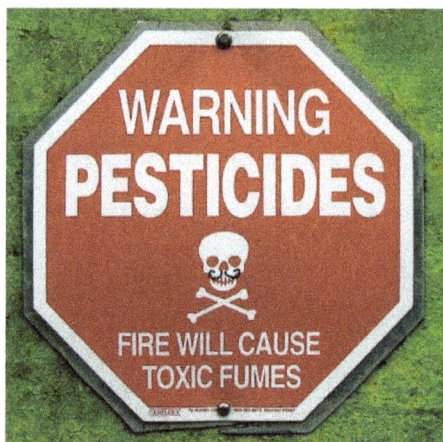

A sign warning about potential pesticide exposure.

The American Academy of Pediatrics recommends limiting exposure of children to pesticides and using safer alternatives:

The World Health Organization and the UN Environment Programme estimate that each year, 3 million workers in agriculture in the developing world experience severe poisoning from pesticides, about 18,000 of whom die. Owing to inadequate regulation and safety precautions, 99% of pesticide related deaths occur in developing countries that account for only 25% of pesticide usage. According to one study, as many as 25 million workers in developing countries may suffer mild pesticide poisoning yearly. There are several careers aside from agriculture that may also put individuals at risk of health effects from pesticide exposure including pet groomers, groundskeepers, and fumigators.

One study found pesticide self-poisoning the method of choice in one third of suicides worldwide, and recommended, among other things, more restrictions on the types of pesticides that are most harmful to humans.

A 2014 epidemiological review found associations between autism and exposure to certain pesticides, but noted that the available evidence was insufficient to conclude that the relationship was causal.

Environmental Effect

Pesticide use raises a number of environmental concerns. Over 98% of sprayed insecticides and 95% of herbicides reach a destination other than their target species, including non-target species, air, water and soil. Pesticide drift occurs when pesticides suspended in the air as particles are carried by wind to other areas, potentially contaminating them. Pesticides are one of the causes of water pollution, and some pesticides are persistent organic pollutants and contribute to soil contamination.

In addition, pesticide use reduces biodiversity, contributes to pollinator decline, destroys habitat (especially for birds), and threatens endangered species. Pests can develop a resistance to the pesticide (pesticide resistance), necessitating a new pesticide. Alternatively a greater dose of the pesticide can be used to counteract the resistance, although this will cause a worsening of the ambient pollution problem.

Since chlorinated hydrocarbon pesticides dissolve in fats and are not excreted, organisms tend to retain them almost indefinitely. Biological magnification is the process whereby these chlorinated hydrocarbons (pesticides) are more concentrated at each level of the food chain. Among marine animals, pesticide concentrations are higher in carnivorous fishes, and even more so in the fish-eating birds and mammals at the top of the ecological pyramid. Global distillation is the process whereby pesticides are transported from warmer to colder regions of the Earth, in particular the Poles and mountain tops. Pesticides that evaporate into the atmosphere at relatively high temperature can be carried considerable distances (thousands of kilometers) by the wind to an area of lower temperature, where they condense and are carried back to the ground in rain or snow.

In order to reduce negative impacts, it is desirable that pesticides be degradable or at least quickly deactivated in the environment. Such loss of activity or toxicity of pesticides is due to both innate chemical properties of the compounds and environmental processes or conditions. For example, the presence of halogens within a chemical structure often slows down degradation in an aerobic environment. Adsorption to soil may retard pesticide movement, but also may reduce bioavailability to microbial degraders.

Economics

Human health and environmental cost from pesticides in the United States is estimated at $9.6 billion offset by about $40 billion in increased agricultural production:

Harm	Annual US cost
Public health	$1.1 billion
Pesticide resistance in pest	$1.5 billion
Crop losses caused by pesticides	$1.4 billion
Bird losses due to pesticides	$2.2 billion
Groundwater contamination	$2.0 billion
Other costs	$1.4 billion
Total costs	**$9.6 billion**

Additional costs include the registration process and the cost of purchasing pesticides. The registration process can take several years to complete (there are 70 different types of field test) and can cost $50–70 million for a single pesticide. Annually the United States spends $10 billion on pesticides.

Alternatives

Alternatives to pesticides are available and include methods of cultivation, use of biological pest controls (such as pheromones and microbial pesticides), genetic engineering, and methods of interfering with insect breeding. Application of composted yard waste has also been used as a way of controlling pests. These methods are becoming increasingly popular and often are safer than traditional chemical pesticides. In addition, EPA is registering reduced-risk conventional pesticides in increasing numbers.

Cultivation practices include polyculture (growing multiple types of plants), crop rotation, planting crops in areas where the pests that damage them do not live, timing planting according to when pests will be least problematic, and use of trap crops that attract pests away from the real crop. In the U.S., farmers have had success controlling insects by spraying with hot water at a cost that is about the same as pesticide spraying.

Release of other organisms that fight the pest is another example of an alternative to pesticide use. These organisms can include natural predators or parasites of the pests. Biological pesticides based on entomopathogenic fungi, bacteria and viruses cause disease in the pest species can also be used.

Interfering with insects' reproduction can be accomplished by sterilizing males of the target species and releasing them, so that they mate with females but do not produce offspring. This technique was first used on the screwworm fly in 1958 and has since been used with the medfly, the tsetse fly, and the gypsy moth. However, this can be a costly, time consuming approach that only works on some types of insects.

Agroecology emphasize nutrient recycling, use of locally available and renewable resources, adaptation to local conditions, utilization of microenvironments, reliance on indigenous knowledge and yield maximization while maintaining soil productivity. Agroecology also emphasizes empowering people and local communities to contribute to development, and encouraging "multi-directional" communications rather than the conventional "top-down" method.

Push Pull Strategy

The term "push-pull" was established in 1987 as an approach for integrated pest management (IPM). This strategy uses a mixture of behavior-modifying stimuli to manipulate the distribution and abundance of insects. "Push" means the insects are repelled or deterred away from whatever resource that is being protected. "Pull" means that certain stimuli (semiochemical stimuli, pheromones, food additives, visual stimuli, genetically altered plants, etc.) are used to attract pests to trap crops where they will be killed. There are numerous different components involved in order to implement a Push-Pull Strategy in IPM.

Many case studies testing the effectiveness of the push-pull approach have been done across the world. The most successful push-pull strategy was developed in Africa for subsistence farming. Another successful case study was performed on the control of *Helicoverpa* in cotton crops in Australia. In Europe, the Middle East, and the United States, push-pull strategies were successfully used in the controlling of *Sitona lineatus* in bean fields.

Some advantages of using the push-pull method are less use of chemical or biological materials and better protection against insect habituation to this control method. Some disadvantages of the push-pull strategy is that if there is a lack of appropriate knowledge of behavioral and chemical ecology of the host-pest interactions then this method becomes unreliable. Furthermore, because the push-pull method is not a very popular method of IPM operational and registration costs are higher.

Effectiveness

Some evidence shows that alternatives to pesticides can be equally effective as the use of chemicals. For example, Sweden has halved its use of pesticides with hardly any reduction in crops.In Indonesia, farmers have reduced pesticide use on rice fields by 65% and experienced a 15% crop increase. A study of Maize fields in northern Florida found that the application of composted yard waste with high carbon to nitrogen ratio to agricultural fields was highly effective at reducing the population of plant-parasitic nematodes and increasing crop yield, with yield increases ranging from 10% to 212%; the observed effects were long-term, often not appearing until the third season of the study.

However, pesticide resistance is increasing. In the 1940s, U.S. farmers lost only 7% of their crops to pests. Since the 1980s, loss has increased to 13%, even though more pesticides are being used.Between 500 and 1,000 insect and weed species have developed pesticide resistance since 1945.

Types

Pesticides are often referred to according to the type of pest they control. Pesticides can also be considered as either biodegradable pesticides, which will be broken down by microbes and other living beings into harmless compounds, or persistent pesticides, which may take months or years before they are broken down: it was the persistence of DDT, for example, which led to its accumulation in the food chain and its killing of birds of prey at the top of the food chain. Another way to think about pesticides is to consider those that are chemical pesticides or are derived from a common source or production method.

Some examples of chemically-related pesticides are:

Organophosphate Pesticides

Organophosphates affect the nervous system by disrupting, acetylcholinesterase activity, the enzyme that regulates acetylcholine, a neurotransmitter. Most organophosphates are insecticides. They were developed during the early 19th century, but their effects on insects, which are similar to their effects on humans, were discovered in 1932. Some are very poisonous. However, they usually are not persistent in the environment.

Carbamate Pesticides

Carbamate pesticides affect the nervous system by disrupting an enzyme that regulates acetylcholine, a neurotransmitter. The enzyme effects are usually reversible. There are several subgroups within the carbamates.

Organochlorine Insecticides

They were commonly used in the past, but many have been removed from the market due to their health and environmental effects and their persistence (e.g., DDT, chlordane, and toxaphene).

Pyrethroid Pesticides

They were developed as a synthetic version of the naturally occurring pesticide pyrethrin, which is found in chrysanthemums. They have been modified to increase their stability in the environment. Some synthetic pyrethroids are toxic to the nervous system.

Sulfonylurea Herbicides

The following sulfonylureas have been commercialized for weed control: amidosulfuron, azimsulfuron, bensulfuron-methyl, chlorimuron-ethyl, ethoxysulfuron, flazasulfuron, flupyrsulfuron-methyl-sodium, halosulfuron-methyl, imazosulfuron, nicosulfuron, oxasulfuron, primisulfuron-methyl, pyrazosulfuron-ethyl, rimsulfuron, sulfometuron-methyl Sulfosulfuron, terbacil, bispyribac-sodium, cyclosulfamuron, and pyrithiobac-sodium. Nicosulfuron, triflusulfuron methyl, and chlorsulfuron are broad-spectrum herbicides that kill plants by inhibiting the enzyme acetolactate synthase. In the 1960s, more than 1 kg/ha (0.89 lb/acre) crop protection chemical was typically applied, while sulfonylureates allow as little as 1% as much material to achieve the same effect.

Biopesticides

Biopesticides are certain types of pesticides derived from such natural materials as animals, plants, bacteria, and certain minerals. For example, canola oil and baking soda have pesticidal applications and are considered biopesticides. Biopesticides fall into three major classes:

- Microbial pesticides which consist of bacteria, entomopathogenic fungi or viruses (and sometimes includes the metabolites that bacteria or fungi produce). Entomopathogenic nematodes are also often classed as microbial pesticides, even though they are multi-cellular.

- Biochemical pesticides or herbal pesticides are naturally occurring substances that control (or monitor in the case of pheromones) pests and microbial diseases.

- Plant-incorporated protectants (PIPs) have genetic material from other species incorporated into their genetic material (*i.e.* GM crops). Their use is controversial, especially in many European countries.

Classified by Type of Pest

Pesticides that are related to the type of pests are:

Type	Action
Algicides	Control algae in lakes, canals, swimming pools, water tanks, and other sites
Antifouling agents	Kill or repel organisms that attach to underwater surfaces, such as boat bottoms
Antimicrobials	Kill microorganisms (such as bacteria and viruses)
Attractants	Attract pests (for example, to lure an insect or rodent to a trap). (However, food is not considered a pesticide when used as an attractant.)
Biopesticides	Biopesticides are certain types of pesticides derived from such natural materials as animals, plants, bacteria, and certain minerals
Biocides	Kill microorganisms
Disinfectants and sanitizers	Kill or inactivate disease-producing microorganisms on inanimate objects
Fungicides	Kill fungi (including blights, mildews, molds, and rusts)

Fumigants	Produce gas or vapor intended to destroy pests in buildings or soil
Herbicides	Kill weeds and other plants that grow where they are not wanted
Insecticides	Kill insects and other arthropods
Miticides	Kill mites that feed on plants and animals
Microbial pesticides	Microorganisms that kill, inhibit, or out compete pests, including insects or other microorganisms
Molluscicides	Kill snails and slugs
Nematicides	Kill nematodes (microscopic, worm-like organisms that feed on plant roots)
Ovicides	Kill eggs of insects and mites
Pheromones	Biochemicals used to disrupt the mating behavior of insects
Repellents	Repel pests, including insects (such as mosquitoes) and birds
Rodenticides	Control mice and other rodents

Further Types of Pesticides

The term pesticide also include these substances:

Defoliants : Cause leaves or other foliage to drop from a plant, usually to facilitate harvest.Desiccants : Promote drying of living tissues, such as unwanted plant tops.Insect growth regulators : Disrupt the molting, maturity from pupal stage to adult, or other life processes of insects.Plant growth regulators : Substances (excluding fertilizers or other plant nutrients) that alter the expected growth, flowering, or reproduction rate of plants.

Regulation

International

In most countries, pesticides must be approved for sale and use by a government agency.

In Europe, recent EU legislation has been approved banning the use of highly toxic pesticides including those that are carcinogenic, mutagenic or toxic to reproduction, those that are endocrine-disrupting, and those that are persistent, bioaccumulative and toxic (PBT) or very persistent and very bioaccumulative (vPvB). Measures were approved to improve the general safety of pesticides across all EU member states.

Though pesticide regulations differ from country to country, pesticides, and products on which they were used are traded across international borders. To deal with inconsistencies in regulations among countries, delegates to a conference of the United Nations Food and Agriculture Organization adopted an International Code of Conduct on the Distribution and Use of Pesticides in 1985 to create voluntary standards of pesticide regulation for different countries. The Code was updated in 1998 and 2002. The FAO claims that the code has raised awareness about pesticide hazards and decreased the number of countries without restrictions on pesticide use.

Three other efforts to improve regulation of international pesticide trade are the United Nations London Guidelines for the Exchange of Information on Chemicals in International Trade and the United Nations Codex Alimentarius Commission. The former seeks to implement procedures for ensuring that prior informed consent exists between countries buying and selling pesticides, while the latter seeks to create uniform standards for maximum levels of pesticide residues among participating countries. Both initiatives operate on a voluntary basis.

Pesticides safety education and pesticide applicator regulation are designed to protect the public from pesticide misuse, but do not eliminate all misuse. Reducing the use of pesticides and choosing less toxic pesticides may reduce risks placed on society and the environment from pesticide use. Integrated pest management, the use of multiple approaches to control pests, is becoming widespread and has been used with success in countries such as Indonesia, China, Bangladesh, the U.S., Australia, and Mexico. IPM attempts to recognize the more widespread impacts of an action on an ecosystem, so that natural balances are not upset. New pesticides are being developed, including biological and botanical derivatives and alternatives that are thought to reduce health and environmental risks. In addition, applicators are being encouraged to consider alternative controls and adopt methods that reduce the use of chemical pesticides.

Pesticides can be created that are targeted to a specific pest's lifecycle, which can be environmentally more friendly. For example, potato cyst nematodes emerge from their protective cysts in response to a chemical excreted by potatoes; they feed on the potatoes and damage the crop. A similar chemical can be applied to fields early, before the potatoes are planted, causing the nematodes to emerge early and starve in the absence of potatoes.

United States

In the United States, the Environmental Protection Agency (EPA) is responsible for regulating pesticides under the Federal Insecticide, Fungicide, and Rodenticide Act (FIFRA) and the Food Quality Protection Act (FQPA). Studies must be conducted to establish the conditions in which the material is safe to use and the effectiveness against the intended pest(s). The EPA regulates pesticides to ensure that these products do not pose adverse effects to humans or the environment. Pesticides produced before November 1984 continue to be reassessed in order to meet the current scientific and regulatory standards. All registered pesticides are reviewed every 15 years to ensure they meet the proper standards. During the registration process, a label is created. The label contains directions for proper use of the material in addition to safety restrictions. Based on acute toxicity, pesticides are assigned to a Toxicity Class.

Some pesticides are considered too hazardous for sale to the general public and are designated restricted use pesticides. Only certified applicators, who have passed an exam, may purchase or supervise the application of restricted use pesticides. Records of sales and use are required to be maintained and may be audited by government agencies charged with the enforcement of pesticide regulations. These records must be made available to employees and state or territorial environmental regulatory agencies.

Preparation for an application of hazardous herbicide in USA.

The EPA regulates pesticides under two main acts, both of which amended by the Food Quality Protection Act of 1996. In addition to the EPA, the United States Department of Agriculture (USDA) and the United States Food and Drug Administration (FDA) set standards for the level of pesticide residue that is allowed on or in crops. The EPA looks at what the potential human health and environmental effects might be associated with the use of the pesticide.

In addition, the U.S. EPA uses the National Research Council's four-step process for human health risk assessment: (1) Hazard Identification, (2) Dose-Response Assessment, (3) Exposure Assessment, and (4) Risk Characterization.

Recently Kaua'i County (Hawai'i) passed Bill No. 2491 to add an article to Chapter 22 of the county's code relating to pesticides and GMOs. The bill strengthens protections of local communities in Kaua'i where many large pesticide companies test their products.

History

Since before 2000 BC, humans have utilized pesticides to protect their crops. The first known pesticide was elemental sulfur dusting used in ancient Sumer about 4,500 years ago in ancient Mesopotamia. The Rig Veda, which is about 4,000 years old, mentions the use of poisonous plants for pest control. By the 15th century, toxic chemicals such as arsenic, mercury, and lead were being applied to crops to kill pests. In the 17th century, nicotine sulfate was extracted from tobacco leaves for use as an insecticide. The 19th century saw the introduction of two more natural pesticides, pyrethrum, which is derived from chrysanthemums, and rotenone, which is derived from the roots of tropical vegetables. Until the 1950s, arsenic-based pesticides were dominant. Paul Müller discovered that DDT was a very effective insecticide. Organochlorines such as DDT were dominant, but they were replaced in the U.S. by organophosphates and carbamates by 1975. Since then, pyrethrin compounds have become the dominant insecticide. Herbicides became common in the 1960s, led by "triazine and other nitrogen-based compounds, carboxylic acids such as 2,4-dichlorophenoxyacetic acid, and glyphosate".

The first legislation providing federal authority for regulating pesticides was enacted in 1910; however, decades later during the 1940s manufacturers began to produce large amounts of synthetic pesticides and their use became widespread. Some sources consider the 1940s and 1950s to have been the start of the "pesticide era." Although the U.S. Environmental Protection Agency was established in 1970 and amendments to the pesticide law in 1972, pesticide use has increased 50-fold since 1950 and 2.3 million tonnes (2.5 million short tons) of industrial pesticides are now used each year. Seventy-five percent of all pesticides in the world are used in developed countries, but use in developing countries is increasing. A study of USA pesticide use trends through 1997 was published in 2003 by the National Science Foundation's Center for Integrated Pest Management.

In the 1960s, it was discovered that DDT was preventing many fish-eating birds from reproducing, which was a serious threat to biodiversity. Rachel Carson wrote the best-selling book *Silent Spring* about biological magnification. The agricultural use of DDT is now banned under the Stockholm Convention on Persistent Organic Pollutants, but it is still used in some developing nations to prevent malaria and other tropical diseases by spraying on interior walls to kill or repel mosquitoes.

Pentachlorobenzene

Pentachlorobenzene (PeCB) is a chemical compound with the molecular formula C_6HCl_5 which is a chlorinated aromatic hydrocarbon. It consists of a benzene ring substituted with five chlorine atoms. PeCB was once used industrially for a variety of uses, but because of environmental concerns there are currently no large scale uses of PeCB. Pentachlorobenzene is a known persistent organic pollutants (POP), classified among the "dirty dozen" and banned globally by the Stockholm Convention on Persistent Organic Pollutants as of 2011.

Production

PeCB can be produced as a byproduct of the manufacture of carbon tetrachloride and benzene. It is extracted by distillation and crystallization. The direct production of pure PeCB is not practical because of the simultaneous production of other chlorinated compounds. Since PeCB is generally produced in small quantities in the chlorination of benzene, it is also contained in other chlorobenzenes (dichlorobenzenes, trichlorobenzenes, etc.)

Today, a majority of the PeCB released into the environment is a result of backyard trash burning and municipal waste incineration.

Uses

PeCB was used as an intermediate in the manufacture of pesticides, particularly the fungicide pentachloronitrobenzene. Pentachloronitrobenzene is now made by the chlorination of nitrobenzene in order to avoid the use of PeCB.

PeCB was a component of a mixture of chlorobenzenes added to products containing polychlorinated biphenyls in order to reduce viscosity. PeCB has also been used as a fire retardant.

Safety and Regulation

PeCB is a persistent organic pollutant, allowing an accumulation in the food chain. Consequently, pentachlorobenzene was added in 2009 to the list of chemical compounds covered by the Stockholm Convention, an international treaty which restricts the production and use of persistent organic pollutants. PeCB has been banned in the European Union since 2002.

PeCB is very toxic to aquatic organisms, and decomposes on heating or on burning with the formation of toxic, corrosive fumes including hydrogen chloride. Combustion of PeCB may also result in the formation of polychlorinated dibenzodioxins ("dioxins") and polychlorinated dibenzofurans.

Hexachlorocyclopentadiene

Hexachlorocyclopentadiene, also known as C-56, is an organochlorine compound that is a precursor to several pesticides. This colourless liquid is an inexpensive reactive diene. Many of its derivatives proved to be highly controversial, as studies showed them to be persistent organic pollutants. Collectively, the pesticides derived from hexachlorocyclopentadiene are called the cyclodienes. An estimated 270,000 tons were produced until 1976, and some are still manufactured by Velsicol Chemical Corporation in the US and by Jiangsu Anpon Electrochemicals Co. in China.

Synthesis and Applications

Hexachlorocyclopentadiene is prepared by chlorination of cyclopentadiene to give 1,1,2,3,4,5-octachlorocyclopentane, which in a second step undergoes dehydrochlorination:

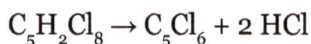

$$C_5H_6 + 6\,Cl_2 \rightarrow C_5H_2Cl_8 + 4\,HCl$$

$$C_5H_2Cl_8 \rightarrow C_5Cl_6 + 2\,HCl$$

Hexachlorocyclopentadiene readily undergoes the Diels-Alder reaction to give a variety of adducts that were commercialized as pesticides. The main derivatives are:

- aldrin from norbornadiene (the related dieldrin is a metabolite of aldrin)

- bromodan from allyl bromide

- chlordane from cyclopentadiene, followed by chlorination

- endrin from acetylene, followed by cyclopentadiene, followed by epoxidation

- heptachlor from cyclopentadiene, followed by monochlorination

- isobenzan from dihydrofuran followed by chlorination

- endosulfan from 1,4-dihydroxy-2-butene, followed by esterification with $SOCl_2$

Additionally hexachlorocyclopentadiene is the precursor to the pesticides mirex and kepone, although these are not classified as cyclodienes.

Regulation

Almost all derivatives have been banned or are under consideration for banning, according to the deliberations of the Stockholm Convention on Persistent Organic Pollutants.

Insect Resistance

In addition to regulatory pressures, these pesticides became less effective owing to genetic mutations of the targeted insects. The number of insects resistant to cyclodienes and lindane approached 300 by 1989.

Chlordane

Chlordane, or chlordan, is an organochlorine compound used as a pesticide. This white solid was sold in the U.S. until 1988 as an insecticide for treating approximately 30 million homes for termites for crops like corn and citrus, and on lawns and domestic gardens. Technical grade chlordane is a complex mixture of over 120 structurally related chemical compounds.

Production, Composition and Uses

Chlordane is one so-called cyclodiene pesticide, meaning that it is derived from hexachlorocyclopentadiene.

Hexachlorocyclopentadiene forms an adduct with cyclopentadiene, and chlorination of this adduct gives predominantly two isomers, α and β, in addition to other products such as *trans*-nonachlor and heptachlor. The β-isomer is popularly known as gamma and is more bioactive. The mixture that is composed of 147 components is called technical chlordane.

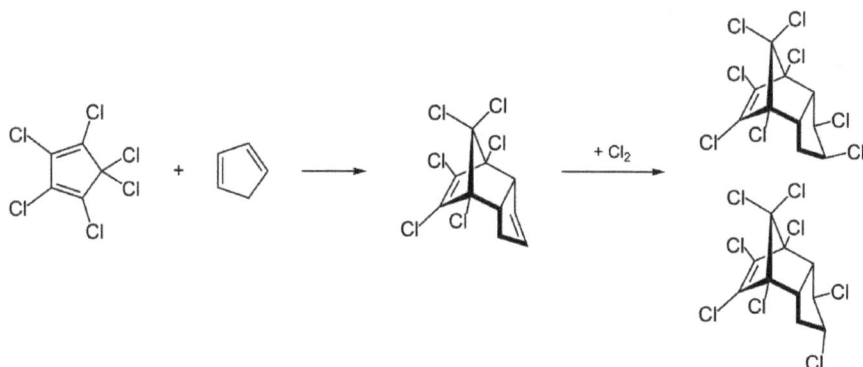

Synthesis of *cis*- (above) and *trans*-chlordane (below)

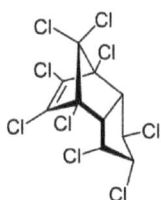

trans-nonachlor

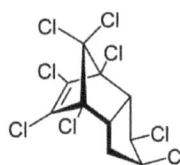

cis-chlordane (also known as α-chlordane
(CAS=5103-71-9))

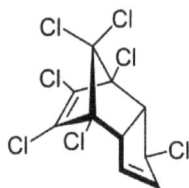

(+)-heptachlor

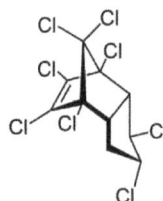

trans-chlordane (also known as γ-chlordane
and beta-chlordane (CAS=5103-74-2))

It was sold in the United States from 1948 to 1988, both as a dust and an emulsified solution.

Because of concern about damage to the environment and harm to human health, the United States Environmental Protection Agency (EPA) banned all uses of chlordane in 1983, except termite control. The EPA banned all uses of chlordane in 1988. The EPA recommends that children should not drink water with more than 60 parts of chlordane per billion parts of drinking water (60 ppb) for longer than 1 day. EPA has set a limit in drinking water of 2 ppb.

Chlordane is very persistent in the environment because it does not break down easily. Recent tests of the air in the residence of U.S. government housing, 32 years after chlordane treatment, showed levels of chlordane and heptachlor 10-15 times the Minimal Risk Levels (20 nanograms/cubic meter of air) published by the Centers for Disease Control. It has an environmental half-life of 10 to 20 years.

Origin, Pathways of Exposure, and Processes of Excretion

In the years 1948–1988 chlordane was a common pesticide for corn and citrus crops, as well as a method of home termite control. Pathways of exposure to chlordane include ingestion of crops grown in chlordane-contaminated soil, inhalation of air near chlordane-treated homes and landfills, and ingestion of high-fat foods such as meat, fish, and dairy, as chlordane builds up in fatty tissue. The United States Environmental Protection Agency reported that over 30 million homes were treated with technical chlordane or technical chlordane with heptachlor. Depending on the site of home treatment, the indoor air levels of chlordane can still exceed the Minimal Risks Levels (MRLs) for both cancer and chronic disease by orders of magnitude. Chlordane is excreted slowly through feces, urine elimination, and through breast milk in nursing mothers. It is able to cross the placenta and become absorbed by developing fetuses in pregnant women. A breakdown product of chlordane, the metabolite oxychlordane, accumulates in blood and adipose tissue with age.

Environmental Impact

Being hydrophobic, chlordane adheres to soil particles and enters groundwater only slowly, owing to its low solubility (0.009 ppm). It degrades only over the course of years. Chlordane bioaccumulates in animals. It is highly toxic to fish, with an LD_{50} of 0.022–0.095 mg/kg (oral).

Two components of the chlordane mixture, *cis*-nonachlor and *trans*-nonachlor, are the main bioaccumulating constituents. *trans*-Nonachlor is more toxic than technical chlordane and *cis*-nonachlor is less toxic. Oxychlordane ($C_{10}H_4Cl_8O$) is the primary metabolite of chlordane.

Chlordane is a known persistent organic pollutants (POP), classified among the "dirty dozen" and banned by the 2001 Stockholm Convention on Persistent Organic Pollutants.

Health Effects

Multiple studies published in the last five years that measured metabolites of chlordane/heptachlor in the blood of U. S. citizens during the U.S. National Health and Examination Surveys (NHANES)(1999-2006) reported that higher concentrations of heptachlor epoxide and oxychlordane increase the risk of cognitive decline, liver damage (liver enzymes), peripheral arterial disease, prostate cancer (trans-nonachlor), type 2 diabetes, and obesity (waist circumference),

In other large epidemiological surveys, higher levels of oxychlordane in both blood and adipose increased the risk of non-Hodgkin lymphoma, and likewise higher concentrations of heptachlor epoxide in brain tissues increase the risk of Parkinson diseases,

Exposure to chlordane metabolites may be associated with testicular cancer. The incidence of seminoma in men with the highest blood levels of *cis*-nonachlor was almost double that of men with the lowest levels. Japanese workers who used chlordane over a long period of time had minor changes in liver function.

Heptachlor and chlordane are some of the most potent carcinogens tested in animal models. No human epidemiological study has been conducted to determine the relationship between levels of chlordane/heptachlor in indoor air and rates of cancer in inhabitants. However, studies have linked chlordane/heptachlor in human tissues with cancers of the breast, prostate, brain, and cancer of blood cells—leukemia and lymphoma. Breathing chlordane in indoor air is the main route of exposure for these levels in human tissues. Currently, USEPA has defined a concentration of 24 nanogram per cubic meter of air (ng/M^3) for chlordane compounds over a 20-year exposure period as the concentration that will increase the probability of cancer by 1 in 1,000,000 persons. This probability of developing cancer increases to 10 in 1,000,000 persons with an exposure of 100 ng/M^3 and 100 in 1,000,000 with an exposure of 1000 ng/M^3.

The non-cancer health effects of chlordane compounds, which include diabetes, insulin resistance, migraines, respiratory infections, immune-system activation, anxiety, depression, blurry vision, confusion, intractable seizures as well as permanent neurological damage, probably affects more people than cancer. Recently, trans-nonachlor and oxychlordane in serum of mothers during gestation has been linked with behaviors associated with autism in offspring at age 4-5. The Agency for Toxic Substances and Disease Registry (ATSDR) has defined a concentration of chlordane compounds of 20 ng/M^3 as the Minimal Risk Level (MRLs). ATSDR defines Minimal Risk Level as an estimate of daily human exposure to a dose of a chemical that is likely to be without an appreciable risk of adverse non-cancerous effects over a specific duration of exposure. Recent results from 8 large epidemiological studies in the United States, using CDC's NHANES data, have consistently shown of all the chemicals found in the blood of Americans, heptachlor epoxides and oxychlordane have the highest associated risk with insulin resistance and diabetes.

Remediation

Chlordane was applied under the home/building during treatment for termites and the half-life can be up to 30 years. Chlordane has a low vapor pressure and volitizes slowly into the air of home/building above. To remove chlordane from indoor air requires either ventilation (Heat Exchange Ventilation) or activated carbon filtration. Chemical remediation of chlordane in soils was attempted by the US Army Corps of Engineers by mixing chlordane with aqueous lime and persulfate. In a phytoremediation study, Kentucky bluegrass and Perennial ryegrass were found to be minimally affected by chlordane, and both were found to take it up into their roots and shoots. Mycoremediation of chlordane in soil have found that contamination levels were reduced. The fungus *Phanerochaete chrysosporium* has been found to reduce concentrations by 21% in water in 30 days and in solids in 60 days.

Herbicide

Herbicide(s), also commonly known as weedkillers, are chemical substances used to control unwanted plants. Selective herbicides control specific weed species, while leaving the desired crop relatively unharmed, while non-selective herbicides (sometimes called "total weedkillers" in commercial products) can be used to clear waste ground, industrial and construction sites, railways and railway embankments as they kill all plant material with which they come into contact. Apart from selective/non-selective, other important distinctions include *persistence* (also known as *residual action*: how long the product stays in place and remains active), *means of uptake* (whether it is absorbed by above-ground foliage only, through the roots, or by other means), and *mechanism of action* (how it works). Historically, products such as common salt and other metal salts were used as herbicides, however these have gradually fallen out of favor and in some countries a number of these are banned due to their persistence in soil, and toxicity and groundwater contamination concerns. Herbicides have also been used in warfare and conflict.

Weeds controlled with herbicide

Modern herbicides are often synthetic mimics of natural plant hormones which interfere with growth of the target plants. The term organic herbicide has come to mean herbicides intended for organic farming; these are often less efficient and more costly than synthetic herbicides and are based on natural materials. Some plants also produce their own natural herbicides, such as the genus *Juglans* (walnuts), or the tree of heaven; such action of natural herbicides, and other related chemical interactions, is called allelopathy. Due to herbicide resistance - a major concern in agriculture - a number of products also combine herbicides with different means of action.

In the US in 2007, about 83% of all herbicide usage, determined by weight applied, was in agriculture. In 2007, world pesticide expenditures totaled about $39.4 billion; herbicides were about 40% of those sales and constituted the biggest portion, followed by insecticides, fungicides, and other types. Smaller quantities are used in forestry, pasture systems, and management of areas set aside as wildlife habitat.

History

Prior to the widespread use of chemical herbicides, cultural controls, such as altering soil pH, salinity, or fertility levels, were used to control weeds. Mechanical control (including tillage) was also (and still is) used to control weeds.

First Herbicides

Although research into chemical herbicides began in the early 20th century, the first major breakthrough was the result of research conducted in both the UK and the US during the Second World War into the potential use of agents as biological weapons. The first modern herbicide, 2,4-D, was first discovered and synthesized by W. G. Templeman at Imperial Chemical Industries. In 1940, he showed that "Growth substances applied appropriately would kill certain broad-leaved weeds in cereals without harming the crops." By 1941, his team succeeded in synthesizing the chemical. In the same year, Pokorny in the US achieved this as well.

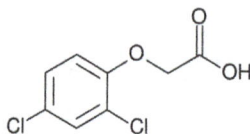

2,4-D, the first chemical herbicide, was discovered during the Second World War.

Independently, a team under Juda Hirsch Quastel, working at the Rothamsted Experimental Station made the same discovery. Quastel was tasked by the Agricultural Research Council (ARC) to discover methods for improving crop yield. By analyzing soil as a dynamic system, rather than an inert substance, he was able to apply techniques such as perfusion. Quastel was able to quantify the influence of various plant hormones, inhibitors and other chemicals on the activity of microorganisms in the soil and assess their direct impact on plant growth. While the full work of the unit remained secret, certain discoveries were developed for commercial use after the war, including the 2,4-D compound.

When it was commercially released in 1946, it triggered a worldwide revolution in agricultural output and became the first successful selective herbicide. It allowed for greatly enhanced weed control in wheat, maize (corn), rice, and similar cereal grass crops, because it kills dicots (broadleaf plants), but not most monocots (grasses). The low cost of 2,4-D has led to continued usage today, and it remains one of the most commonly used herbicides in the world. Like other acid herbicides, current formulations use either an amine salt (often trimethylamine) or one of many esters of the parent compound. These are easier to handle than the acid.

Further Discoveries

The triazine family of herbicides, which includes atrazine, were introduced in the 1950s; they have the current distinction of being the herbicide family of greatest concern regarding groundwater contamination. Atrazine does not break down readily (within a few weeks) after being applied to soils of above neutral pH. Under alkaline soil conditions, atrazine may be carried into the soil profile as far as the water table by soil water following rainfall causing the aforementioned contamination. Atrazine is thus said to have "carryover", a generally undesirable property for herbicides.

Glyphosate (Roundup) was introduced in 1974 for nonselective weed control. Following the development of glyphosate-resistant crop plants, it is now used very extensively for selective weed control in growing crops. The pairing of the herbicide with the resistant seed contributed to the consolidation of the seed and chemistry industry in the late 1990s.

Many modern chemical herbicides used in agriculture and gardening are specifically formulated to decompose within a short period after application. This is desirable, as it allows crops and plants to be planted afterwards, which could otherwise be affected by the herbicide. However, herbicides with low residual activity (i.e., that decompose quickly) often do not provide season-long weed control and do not ensure that weed roots are killed beneath construction and paving (and cannot emerge destructively in years to come), therefore there remains a role for weedkiller with high levels of persistence in the soil.

Terminology

Herbicides are classified/grouped in various ways e.g. according to the activity, timing of application, method of application, mechanism of action, chemical family. This gives rise to a considerable level of terminology related to herbicides and their use.

Intended Outcome

- Control is the destruction of unwanted weeds, or the damage of them to the point where they are no longer competitive with the crop.

- Suppression is incomplete control still providing some economic benefit, such as reduced competition with the crop.

- Crop safety, for selective herbicides, is the relative absence of damage or stress to the crop. Most selective herbicides cause some visible stress to crop plants.

- Defoliant, similar to herbicides, but designed to remove foliage (leaves) rather than kill the plant.

Selectivity (all Plants or Specific Plants)

- Selective herbicides: They control or suppress certain plants without affecting the growth of other plants species. Selectivity may be due to translocation, differential absorption, physical (morphological) or physiological differences between plant species. 2,4-D, mecoprop, dicamba control many broadleaf weeds but remain ineffective against turfgrasses.

- Non-selective herbicides: These herbicides are not specific in acting against certain plant species and control all plant material with which they come into contact. They are used to clear industrial sites, waste ground, railways and railway embankments. Paraquat, glufosinate, glyphosate are non-selective herbicides.

Timing of Application

- Preplant: Preplant herbicides are nonselective herbicides applied to soil before planting. Some preplant herbicides may be mechanically incorporated into the soil. The objective for incorporation is to prevent dissipation through photodecomposition and/or volatility. The herbicides kill weeds as they grow through the herbicide treated zone. Volatile herbicides have to be incorporated into the soil before planting the pasture. Agricultural crops grown in soil treated with a preplant herbicide include tomatoes, corn, soybeans and strawberries. Soil fumigants like metam-sodium and dazomet are in use as preplant herbicides.

- Preemergence: Preemergence herbicides are applied before the weed seedlings emerge through the soil surface. Herbicides do not prevent weeds from germinating but they kill weeds as they grow through the herbicide treated zone by affecting the cell division in the emerging seedling. Dithopyr and pendimethalin are preemergence herbicides. Weeds that have already emerged before application or activation are not affected by pre-herbicides as their primary growing point escapes the treatment.

- Postemergence: These herbicides are applied after weed seedlings have emerged through the soil surface. They can be foliar or root absorbed, selective or non-selective, contact or systemic. Application of these herbicides is avoided during rain because the problem of being washed off to the soil makes it ineffective. 2,4-D is a selective, systemic, foliar absorbed postemergence herbicide.

Method of Application

- Soil applied: Herbicides applied to the soil are usually taken up by the root or shoot of the emerging seedlings and are used as preplant or preemergence treatment. Several factors influence the effectiveness of soil-applied herbicides. Weeds absorb herbicides by both passive and active mechanism. Herbicide adsorption to soil colloids or organic matter often reduces its amount available for weed absorption. Positioning of herbicide in correct layer of soil is very important, which can be achieved mechanically and by rainfall. Herbicides on the soil surface are subjected to several processes that reduce their availability. Volatility and photolysis are two common processes that reduce the availability of herbicides. Many soil applied herbicides are absorbed through plant shoots while they are still underground leading to their death or injury. EPTC and trifluralin are soil applied herbicides.

- Foliar applied: These are applied to portion of the plant above the ground and are absorbed by exposed tissues. These are generally postemergence herbicides and can either be translocated (systemic) throughout the plant or remain at specific site (contact). External barriers of plants like cuticle, waxes, cell wall etc. affect herbicide absorption and action. Glyphosate, 2,4-D and dicamba are foliar applied herbicide.

Persistence

- Residual activity: A herbicide is described as having low residual activity if it is neutralized within a short time of application (within a few weeks or months) - typically this is due to rainfall, or by reactions in the soil. A herbicide described as having high residual activity will remains potent for a long term in the soil. For some compounds, the residual activity can leave the ground almost permanently barren.

Mechanism of Action

Herbicides are often classified according to their site of action, because as a general rule, herbicides within the same site of action class will produce similar symptoms on susceptible plants. Classification based on site of action of herbicide is comparatively better as herbicide resistance management can be handled more properly and effectively. Classification by mechanism of action (MOA) indicates the first enzyme, protein, or biochemical step affected in the plant following application.

List of Mechanisms Found in Modern Herbicides

- ACCase inhibitors compounds kill grasses. Acetyl coenzyme A carboxylase (ACCase) is part of the first step of lipid synthesis. Thus, ACCase inhibitors affect cell membrane production in the meristems of the grass plant. The ACCases of grasses are sensitive to these herbicides, whereas the ACCases of dicot plants are not.

- ALS inhibitors: the acetolactate synthase (ALS) enzyme (also known as acetohydroxyacid synthase, or AHAS) is the first step in the synthesis of the branched-chain amino acids (valine, leucine, and isoleucine). These herbicides slowly starve affected plants of these amino acids, which eventually leads to inhibition of DNA synthesis. They affect grasses and dicots alike. The ALS inhibitor family includes various sulfonylureas (such as Flazasulfuron and Metsulfuron-methyl), imidazolinones, triazolopyrimidines, pyrimidinyl oxybenzoates, and sulfonylamino carbonyl triazolinones. The ALS biological pathway exists only in plants and not animals, thus making the ALS-inhibitors among the safest herbicides.

- EPSPS inhibitors: The enolpyruvylshikimate 3-phosphate synthase enzyme EPSPS is used in the synthesis of the amino acids tryptophan, phenylalanine and tyrosine. They affect grasses and dicots alike. Glyphosate (Roundup) is a systemic EPSPS inhibitor inactivated by soil contact.

- Synthetic auxins inaugurated the era of organic herbicides. They were discovered in the 1940s after a long study of the plant growth regulator auxin. Synthetic auxins mimic this plant hormone. They have several points of action on the cell membrane, and are effective in the control of dicot plants. 2,4-D is a synthetic auxin herbicide.

- Photosystem II inhibitors reduce electron flow from water to $NADPH2+$ at the photochemical step in photosynthesis. They bind to the Qb site on the D1 protein, and prevent quinone from binding to this site. Therefore, this group of compounds causes electrons to accumulate on chlorophyll molecules. As a consequence, oxidation reactions in excess of those normally tolerated by the cell occur, and the plant dies. The triazine herbicides (including atrazine) and urea derivatives (diuron) are photosystem II inhibitors.

- Photosystem I inhibitors steal electrons from the normal pathway through FeS to Fdx to NADP leading to direct discharge of electrons on oxygen. As a result, reactive oxygen species are produced and oxidation reactions in excess of those normally tolerated by the cell occur, leading to plant death. Bipyridinium herbicides (such as diquat and paraquat) inhibit the Fe-S – Fdx step of that chain, while diphenyl ether herbicides (such as nitrofen, nitrofluorfen, and acifluorfen) inhibit the Fdx – NADP step.

- HPPD inhibitors inhibit 4-Hydroxyphenylpyruvate dioxygenase, which are involved in tyrosine breakdown. Tyrosine breakdown products are used by plants to make carotenoids, which protect chlorophyll in plants from being destroyed by sunlight. If this happens, the plants turn white due to complete loss of chlorophyll, and the plants die. Mesotrione and sulcotrione are herbicides in this class; a drug, nitisinone, was discovered in the course of developing this class of herbicides.

Herbicide Group (Labeling)

One of the most important methods for preventing, delaying, or managing resistance is to reduce the reliance on a single herbicide mode of action. To do this, farmers must know the mode of action for the herbicides they intend to use, but the relatively complex nature of plant biochemistry makes this difficult to determine. Attempts were made to simplify the understanding of herbicide mode of action by developing a classification system that grouped herbicides by mode of action. Eventually the Herbicide Resistance Action Committee (HRAC) and the Weed Science Society of America (WSSA) developed a classification system. The WSSA and HRAC systems differ in the group designation. Groups in the WSSA and the HRAC systems are designated by numbers and letters, respectively. The goal for adding the "Group" classification and mode of action to the herbicide product label is to provide a simple and practical approach to deliver the information to users. This information will make it easier to develop educational material that is consistent and effective. It should increase user's awareness of herbicide mode of action and provide more accurate recommendations for resistance management. Another goal is to make it easier for users to keep records on which herbicide mode of actions are being used on a particular field from year to year.

Chemical Family

Detailed investigations on chemical structure of the active ingredients of the registered herbicides showed that some moieties (moiety is a part of a molecule that may include either whole functional groups or parts of functional groups as substructures; a functional group has similar chemical properties whenever it occurs in different compounds) have the same mechanisms of action. According to Forouzesh *et al.* 2015, these moieties have been assigned to the names of chemical families and active ingredients are then classified within the chemical families accordingly. Knowing about herbicide chemical family grouping could serve as a short-term strategy for managing resistance to site of action.

Use and Application

Most herbicides are applied as water-based sprays using ground equipment. Ground equipment varies in design, but large areas can be sprayed using self-propelled sprayers equipped with long booms, of 60 to 120 feet (18 to 37 m) with spray nozzles spaced every 20–30 inches (510–760 mm) apart. Towed, handheld, and even horse-drawn sprayers are also used. On large areas, herbicides may also at times be applied aerially using helicopters or airplanes, or through irrigation systems (known as chemigation).

Herbicides being sprayed from the spray arms of a tractor in North Dakota.

A further method of herbicide application developed around 2010, involves ridding the soil of its active weed seed bank rather than just killing the weed. This can successfully treat annual plants but not perennials. Researchers at the Agricultural Research Service found that the application of herbicides to fields late in the weeds' growing season greatly reduces their seed production, and therefore fewer weeds will return the following season. Because most weeds are annuals, their seeds will only survive in soil for a year or two, so this method will be able to destroy such weeds after a few years of herbicide application.

Weed-wiping may also be used, where a wick wetted with herbicide is suspended from a boom and dragged or rolled across the tops of the taller weed plants. This allows treatment of taller grassland weeds by direct contact without affecting related but desirable shorter plants in the grassland sward beneath. The method has the benefit of avoiding spray drift. In Wales, a scheme offering free weed-wiper hire was launched in 2015 in an effort to reduce the levels of MCPA in water courses.

Misuse and Misapplication

Herbicide volatilisation or spray drift may result in herbicide affecting neighboring fields or plants, particularly in windy conditions. Sometimes, the wrong field or plants may be sprayed due to error.

Use Politically, Militarily, and in Conflict

Health and Environmental Effects

Herbicides have widely variable toxicity in addition to acute toxicity from occupational exposure levels.

Some herbicides cause a range of health effects ranging from skin rashes to death. The pathway of attack can arise from intentional or unintentional direct consumption, improper application resulting in the herbicide coming into direct contact with people or wildlife, inhalation of aerial sprays, or food consumption prior to the labeled preharvest interval. Under some conditions, certain herbicides can be transported via leaching or surface runoff to contaminate groundwater or distant surface water sources. Generally, the conditions that promote herbicide transport include intense storm events (particularly shortly after application) and soils with limited capacity to adsorb or retain the herbicides. Herbicide properties that increase likelihood of transport include persistence (resistance to degradation) and high water solubility.

Phenoxy herbicides are often contaminated with dioxins such as TCDD; research has suggested such contamination results in a small rise in cancer risk after occupational exposure to these herbicides. Triazine exposure has been implicated in a likely relationship to increased risk of breast cancer, although a causal relationship remains unclear.

Herbicide manufacturers have at times made false or misleading claims about the safety of their products. Chemical manufacturer Monsanto Company agreed to change its advertising after pressure from New York attorney general Dennis Vacco; Vacco complained about misleading claims that its spray-on glyphosate-based herbicides, including Roundup, were safer than table salt and "practically non-toxic" to mammals, birds, and fish (though proof that this was ever said is hard to find). Roundup is toxic and has resulted in death after being ingested in quantities ranging from 85 to 200 ml, although it has also been ingested in quantities as large as 500 ml with only mild or moderate symptoms. The manufacturer of Tordon 101 (Dow AgroSciences, owned by the Dow Chemical Company) has claimed Tordon 101 has no effects on animals and insects, in spite of evidence of strong carcinogenic activity of the active ingredient Picloram in studies on rats.

The risk of Parkinson's disease has been shown to increase with occupational exposure to herbicides and pesticides. The herbicide paraquat is suspected to be one such factor.

All commercially sold, organic and nonorganic herbicides must be extensively tested prior to approval for sale and labeling by the Environmental Protection Agency. However, because of the large number of herbicides in use, concern regarding health effects is significant. In addition to health effects caused by herbicides themselves, commercial herbicide mixtures often contain other chemicals, including inactive ingredients, which have negative impacts on human health.

Ecological Effects

Commercial herbicide use generally has negative impacts on bird populations, although the impacts are highly variable and often require field studies to predict accurately. Laboratory studies have at times overestimated negative impacts on birds due to toxicity, predicting serious problems that were not observed in the field. Most observed effects are due not to toxicity, but to habitat changes and the decreases in abundance of species on which birds rely for food or shelter. Herbicide use in silviculture, used to favor certain types of growth following clearcutting, can cause significant drops in bird populations. Even when herbicides which have low toxicity to birds are used, they decrease the abundance of many types of vegetation on which the birds rely. Herbicide use in agriculture in Britain has been linked to a decline in seed-eating bird species which rely on the weeds killed by the herbicides. Heavy use of herbicides in neotropical agricultural areas has been one of many factors implicated in limiting the usefulness of such agricultural land for wintering migratory birds.

Frog populations may be affected negatively by the use of herbicides as well. While some studies have shown that atrazine may be a teratogen, causing demasculinization in male frogs, the U.S. Environmental Protection Agency (EPA) and its independent Scientific Advisory Panel (SAP) examined all available studies on this topic and concluded that "atrazine does not adversely affect amphibian gonadal development based on a review of laboratory and field studies."

Scientific Uncertainty of Full Extent of Herbicide Effects

The health and environmental effects of many herbicides is unknown, and even the scientific community often disagrees on the risk. For example, a 1995 panel of 13 scientists reviewing studies on the carcinogenicity of 2,4-D had divided opinions on the likelihood 2,4-D causes cancer in humans. As of 1992, studies on phenoxy herbicides were too few to accurately assess the risk of many types of cancer from these herbicides, even though evidence was stronger that exposure to these herbicides is associated with increased risk of soft tissue sarcoma and non-Hodgkin lymphoma. Furthermore, there is some suggestion that herbicides can play a role in sex reversal of certain organisms that experience temperature-dependent sex determination, which could theoretically alter sex ratios.

Resistance

Weed resistance to herbicides has become a major concern in crop production worldwide. Resistance to herbicides is often attributed to lack of rotational programmes of herbicides and to continuous applications of herbicides with the same sites of action. Thus, a true understanding of the sites of action of herbicides is essential for strategic planning of herbicide-based weed control.

Plants have developed resistance to atrazine and to ALS-inhibitors, and more recently, to glyphosate herbicides. Marestail is one weed that has developed glyphosate resistance. Glyphosate-resistant weeds are present in the vast majority of soybean, cotton and corn farms in some U.S. states. Weeds that can resist multiple other herbicides are spreading. Few new herbicides are near commercialization, and none with a molecular mode of action for which there is no resistance. Because most herbicides could not kill all weeds, farmers rotated crops and herbicides to stop resistant weeds. During its initial years, glyphosate was not subject to resistance and allowed farmers to reduce the use of rotation.

A family of weeds that includes waterhemp (Amaranthus rudis) is the largest concern. A 2008-9 survey of 144 populations of waterhemp in 41 Missouri counties revealed glyphosate resistance in 69%. Weeds from some 500 sites throughout Iowa in 2011 and 2012 revealed glyphosate resistance in approximately 64% of waterhemp samples. The use of other killers to target "residual" weeds has become common, and may be sufficient to have stopped the spread of resistance From 2005 through 2010 researchers discovered 13 different weed species that had developed resistance to glyphosate. But since then only two more have been discovered. Weeds resistant to multiple herbicides with completely different biological action modes are on the rise. In Missouri, 43% of samples were resistant to two different herbicides; 6% resisted three; and 0.5% resisted four. In Iowa 89% of waterhemp samples resist two or more herbicides, 25% resist three, and 10% resist five.

For southern cotton, herbicide costs has climbed from between $50 and $75 per hectare a few years ago to about $370 per hectare in 2013. Resistance is contributing to a massive shift away from growing cotton; over the past few years, the area planted with cotton has declined by 70% in Arkansas and by 60% in Tennessee. For soybeans in Illinois, costs have risen from about $25 to $160 per hectare.

Dow, Bayer CropScience, Syngenta and Monsanto are all developing seed varieties resistant to herbicides other than glyphosate, which will make it easier for farmers to use alternative weed killers. Even though weeds have already evolved some resistance to those herbicides, Powles says the new seed-and-herbicide combos should work well if used with proper rotation.

Biochemistry of Resistance

Resistance to herbicides can be based on one of the following biochemical mechanisms:

- Target-site resistance: This is due to a reduced (or even lost) ability of the herbicide to bind to its target protein. The effect usually relates to an enzyme with a crucial function in a metabolic pathway, or to a component of an electron-transport system. Target-site resistance may also be caused by an overexpression of the target enzyme (via gene amplification or changes in a gene promoter).

- Non-target-site resistance: This is caused by mechanisms that reduce the amount of herbicidal active compound reaching the target site. One important mechanism is an enhanced metabolic detoxification of the herbicide in the weed, which leads to insufficient amounts of the active substance reaching the target site. A reduced uptake and translocation, or sequestration of the herbicide, may also result in an insufficient herbicide transport to the target site.

- Cross-resistance: In this case, a single resistance mechanism causes resistance to several herbicides. The term target-site cross-resistance is used when the herbicides bind to the same target site, whereas non-target-site cross-resistance is due to a single non-target-site mechanism (e.g., enhanced metabolic detoxification) that entails resistance across herbicides with different sites of action.

- Multiple resistance: In this situation, two or more resistance mechanisms are present within individual plants, or within a plant population.

Resistance Management

Worldwide experience has been that farmers tend to do little to prevent herbicide resistance developing, and only take action when it is a problem on their own farm or neighbor's. Careful observation is important so that any reduction in herbicide efficacy can be detected. This may indicate evolving resistance. It is vital that resistance is detected at an early stage as if it becomes an acute, whole-farm problem, options are more limited and greater expense is almost inevitable. Table 1 lists factors which enable the risk of resistance to be assessed. An essential pre-requisite for confirmation of resistance is a good diagnostic test.

Ideally this should be rapid, accurate, cheap and accessible. Many diagnostic tests have been developed, including glasshouse pot assays, petri dish assays and chlorophyll fluorescence. A key component of such tests is that the response of the suspect population to a herbicide can be compared with that of known susceptible and resistant standards under controlled conditions. Most cases of herbicide resistance are a consequence of the repeated use of herbicides, often in association with crop monoculture and reduced cultivation practices. It is necessary, therefore, to modify these practices in order to prevent or delay the onset of resistance or to control existing resistant populations. A key objective should be the reduction in selection pressure. An integrated weed management (IWM) approach is required, in which as many tactics as possible are used to combat weeds. In this way, less reliance is placed on herbicides and so selection pressure should be reduced.

Optimising herbicide input to the economic threshold level should avoid the unnecessary use of herbicides and reduce selection pressure. Herbicides should be used to their greatest potential by ensuring that the timing, dose, application method, soil and climatic conditions are optimal for good activity. In the UK, partially resistant grass weeds such as *Alopecurus myosuroides* (blackgrass) and *Avena* spp. (wild oat) can often be controlled adequately when herbicides are applied at the 2-3 leaf stage, whereas later applications at the 2-3 tiller stage can fail badly. Patch spraying, or applying herbicide to only the badly infested areas of fields, is another means of reducing total herbicide use.

Table 1. Agronomic factors influencing the risk of herbicide resistance development

Factor	Low risk	High risk
Cropping system	Good rotation	Crop monoculture
Cultivation system	Annual ploughing	Continuous minimum tillage
Weed control	Cultural only	Herbicide only
Herbicide use	Many modes of action	Single modes of action
Control in previous years	Excellent	Poor
Weed infestation	Low	High
Resistance in vicinity	Unknown	Common

Approaches to Treating Resistant Weeds

Alternative Herbicides

When resistance is first suspected or confirmed, the efficacy of alternatives is likely to be the first consideration. The use of alternative herbicides which remain effective on resistant populations can be a successful strategy, at least in the short term. The effectiveness of alternative herbicides will be highly dependent on the extent of cross-resistance. If there is resistance to a single group of herbicides, then the use of herbicides from other groups may provide a simple and effective solution, at least in the short term. For example, many triazine-resistant weeds have been readily controlled by the use of alternative herbicides such as dicamba or glyphosate. If resistance extends to more than one herbicide group, then choices are more limited. It should not be assumed that resistance will automatically extend to all herbicides with the same mode of action, although it is wise to assume this until proved otherwise. In many weeds the degree of cross-resistance between the five groups of ALS inhibitors varies considerably. Much will depend on the resistance mechanisms present, and it should not be assumed that these will necessarily be the same in different populations of the same species. These differences are due, at least in part, to the existence of different mutations conferring target site resistance. Consequently, selection for different mutations may result in different patterns of cross-resistance. Enhanced metabolism can affect even closely related herbicides to differing degrees. For example, populations of *Alopecurus myosuroides* (blackgrass) with an enhanced metabolism mechanism show resistance to pendimethalin but not to trifluralin, despite both being dinitroanilines. This is due to differences in the vulnerability of these two herbicides to oxidative metabolism. Consequently, care is needed when trying to predict the efficacy of alternative herbicides.

Mixtures and Sequences

The use of two or more herbicides which have differing modes of action can reduce the selection for resistant genotypes. Ideally, each component in a mixture should:

- Be active at different target sites
- Have a high level of efficacy
- Be detoxified by different biochemical pathways
- Have similar persistence in the soil (if it is a residual herbicide)
- Exert negative cross-resistance
- Synergise the activity of the other component

No mixture is likely to have all these attributes, but the first two listed are the most important. There is a risk that mixtures will select for resistance to both components in the longer term. One practical advantage of sequences of two herbicides compared with mixtures is that a better appraisal of the efficacy of each herbicide component is possible, provided that sufficient time elapses between each application. A disadvantage with sequences is that two separate applications have to be made and it is possible that the later application will be less effective on weeds surviving the first application. If these are resistant, then the second herbicide in the sequence may increase selection for resistant individuals by killing the susceptible plants which were damaged but not killed by the first application, but allowing the larger, less affected, resistant plants to survive. This has been cited as one reason why ALS-resistant *Stellaria media* has evolved in Scotland recently (2000), despite the regular use of a sequence incorporating mecoprop, a herbicide with a different mode of action.

Herbicide Rotations

Rotation of herbicides from different chemical groups in successive years should reduce selection for resistance. This is a key element in most resistance prevention programmes. The value of this approach depends on the extent of cross-resistance, and whether multiple resistance occurs owing to the presence of several different resistance mechanisms. A practical problem can be the lack of awareness by farmers of the different groups of herbicides that exist. In Australia a scheme has been introduced in which identifying letters are included on the product label as a means of enabling farmers to distinguish products with different modes of action.

Farming Practices and Resistance: a Case Study

Herbicide resistance became a critical problem in Australian agriculture, after many Australian sheep farmers began to exclusively grow wheat in their pastures in the 1970s. Introduced varieties of ryegrass, while good for grazing sheep, compete intensely with wheat. Ryegrasses produce so many seeds that, if left unchecked, they can completely choke a field. Herbicides provided excellent control, while reducing soil disrupting because of less need to plough. Within little more than a decade, ryegrass and other weeds began to develop resistance. In response Australian farmers changed methods. By 1983, patches of ryegrass had become immune to Hoegrass, a family of herbicides that inhibit an enzyme called acetyl coenzyme A carboxylase.

Ryegrass populations were large, and had substantial genetic diversity, because farmers had planted many varieties. Ryegrass is cross-pollinated by wind, so genes shuffle frequently. To control its distribution farmers sprayed inexpensive Hoegrass, creating selection pressure. In addition, farmers sometimes diluted the herbicide in order to save money, which allowed some plants to survive application. When resistance appeared farmers turned to a group of herbicides that block acetolactate synthase. Once again, ryegrass in Australia evolved a kind of "cross-resistance" that allowed it to rapidly break down a variety of herbicides. Four classes of herbicides become ineffective within a few years. In 2013 only two herbicide classes, called Photosystem II and long-chain fatty acid inhibitors, were effective against ryegrass.

List of Common Herbicides

Synthetic Herbicides

- 2,4-D is a broadleaf herbicide in the phenoxy group used in turf and no-till field crop production. Now, it is mainly used in a blend with other herbicides to allow lower rates of herbicides to be used; it is the most widely used herbicide in the world, and third most commonly used in the United States. It is an example of synthetic auxin (plant hormone).

- Aminopyralid is a broadleaf herbicide in the pyridine group, used to control weeds on grassland, such as docks, thistles and nettles. It is notorious for its ability to persist in compost.

- Atrazine, a triazine herbicide, is used in corn and sorghum for control of broadleaf weeds and grasses. Still used because of its low cost and because it works well on a broad spectrum of weeds common in the US corn belt, atrazine is commonly used with other herbicides to reduce the overall rate of atrazine and to lower the potential for groundwater contamination; it is a photosystem II inhibitor.

- Clopyralid is a broadleaf herbicide in the pyridine group, used mainly in turf, rangeland, and for control of noxious thistles. Notorious for its ability to persist in compost, it is another example of synthetic auxin.

- Dicamba, a postemergent broadleaf herbicide with some soil activity, is used on turf and field corn. It is another example of a synthetic auxin.

- Glufosinate ammonium, a broad-spectrum contact herbicide, is used to control weeds after the crop emerges or for total vegetation control on land not used for cultivation.

- Fluazifop (Fuselade Forte), a post emergence, foliar absorbed, translocated grass-selective herbicide with little residual action. It is used on a very wide range of broad leaved crops for control of annual and perennial grasses.

- Fluroxypyr, a systemic, selective herbicide, is used for the control of broadleaved weeds in small grain cereals, maize, pastures, rangeland and turf. It is a synthetic auxin. In cereal growing, fluroxypyr's key importance is control of cleavers, *Galium aparine*. Other key broadleaf weeds are also controlled.

- Glyphosate, a systemic nonselective herbicide, is used in no-till burndown and for weed control in crops genetically modified to resist its effects. It is an example of an EPSPs inhibitor.

- Imazapyr a nonselective herbicide, is used for the control of a broad range of weeds, including terrestrial annual and perennial grasses and broadleaf herbs, woody species, and riparian and emergent aquatic species.

- Imazapic, a selective herbicide for both the pre- and postemergent control of some annual and perennial grasses and some broadleaf weeds, kills plants by inhibiting the production of branched chain amino acids (valine, leucine, and isoleucine), which are necessary for protein synthesis and cell growth.

- Imazamox, an imidazolinone manufactured by BASF for postemergence application that is an acetolactate synthase (ALS) inhibitor. Sold under trade names Raptor, Beyond, and Clearcast.

- Linuron is a nonselective herbicide used in the control of grasses and broadleaf weeds. It works by inhibiting photosynthesis.

- MCPA (2-methyl-4-chlorophenoxyacetic acid) is a phenoxy herbicide selective for broadleaf plants and widely used in cereals and pasture.

- Metolachlor is a pre-emergent herbicide widely used for control of annual grasses in corn and sorghum; it has displaced some of the atrazine in these uses.

- Paraquat is a nonselective contact herbicide used for no-till burndown and in aerial destruction of marijuana and coca plantings. It is more acutely toxic to people than any other herbicide in widespread commercial use.

- Pendimethalin, a pre-emergent herbicide, is widely used to control annual grasses and some broad-leaf weeds in a wide range of crops, including corn, soybeans, wheat, cotton, many tree and vine crops, and many turfgrass species.

- Picloram, a pyridine herbicide, mainly is used to control unwanted trees in pastures and edges of fields. It is another synthetic auxin.

- Sodium chlorate *(disused/banned in some countries)*, a nonselective herbicide, is considered phytotoxic to all green plant parts. It can also kill through root absorption.

- Triclopyr, a systemic, foliar herbicide in the pyridine group, is used to control broadleaf weeds while leaving grasses and conifers unaffected.

- Several sulfonylureas, including Flazasulfuron and Metsulfuron-methyl, which act as ALS inhibitors and in some cases are taken up from the soil via the roots.

Organic Herbicides

Recently, the term "organic" has come to imply products used in organic farming. Under this definition, an organic herbicide is one that can be used in a farming enterprise that has been classified as organic. Commercially sold organic herbicides are expensive and may not be affordable for commercial farming. Depending on the application, they may be less effective than synthetic herbicides and are generally used along with cultural and mechanical weed control practices.

Homemade organic herbicides include:

- Corn gluten meal (CGM) is a natural pre-emergence weed control used in turfgrass, which reduces germination of many broadleaf and grass weeds.

- Vinegar is effective for 5–20% solutions of acetic acid, with higher concentrations most effective, but it mainly destroys surface growth, so respraying to treat regrowth is needed. Resistant plants generally succumb when weakened by respraying.

- Steam has been applied commercially, but is now considered uneconomical and inadequate. It controls surface growth but not underground growth and so respraying to treat regrowth of perennials is needed.

- Flame is considered more effective than steam, but suffers from the same difficulties.

- D-limonene (citrus oil) is a natural degreasing agent that strips the waxy skin or cuticle from weeds, causing dehydration and ultimately death.

- Saltwater or salt applied in appropriate strengths to the rootzone will kill most plants.

- Monocerin produced by certain fungi will kill certain weeds such as Johnson grass.

Of Historical Interest and other

- 2,4,5-Trichlorophenoxyacetic acid (2,4,5-T) was a widely used broadleaf herbicide until being phased out starting in the late 1970s. While 2,4,5-T itself is of only moderate toxicity, the manufacturing process for 2,4,5-T contaminates this chemical with trace amounts of 2,3,7,8-tetrachlorodibenzo-p-dioxin (TCDD). TCDD is extremely toxic to humans. With proper temperature control during production of 2,4,5-T, TCDD levels can be held to about .005 ppm. Before the TCDD risk was well understood, early production facilities lacked proper temperature controls. Individual batches tested later were found to have as much as 60 ppm of TCDD. 2,4,5-T was withdrawn from use in the USA in 1983, at a time of heightened public sensitivity about chemical hazards in the environment. Public concern about dioxins was high, and production and use of other (non-herbicide) chemicals potentially containing TCDD contamination was also withdrawn. These included pentachlorophenol (a wood preservative) and PCBs (mainly used as stabilizing agents in transformer oil). Some feel that the 2,4,5-T withdrawal was not based on sound science. 2,4,5-T has since largely been replaced by dicamba and triclopyr.

- Agent Orange was a herbicide blend used by the British military during the Malayan Emergency and the U.S. military during the Vietnam War between January 1965 and April 1970 as a defoliant. It was a 50/50 mixture of the n-butyl esters of 2,4,5-T and 2,4-D. Because of TCDD contamination in the 2,4,5-T component, it has been blamed for serious illnesses in many people who were exposed to it. However, research on populations exposed to its dioxin contaminant have been inconsistent and inconclusive.

- Diesel, and other heavy oil derivatives, are known to be informally used at times, but are usually banned for this purpose.

References

- Metcalf, Robert L. (2002). "Ullmann's Encyclopedia of Industrial Chemistry". Ullmann's Encyclopedia of Industrial Chemistry. Wiley-VCH. doi:10.1002/14356007.a14_263. ISBN 3527306730.

- Zitko, Vladimir (2003). Persistent Organic Pollutants (PDF). Berlin Heidelberg: Springer-Verlag. pp. 47–90. ISBN 978-3-540-43728-4. Retrieved 25 March 2015.

- Esch, G. T. van; Heemstra-Lequin, E. A. H. van (1992). Endrin (PDF). Geneva: World Health Organization. pp. 1–234. ISBN 92 4 157130 6. Retrieved 29 March 2015.

- Brandenberger, Hans; Maes, Robert A. A. (1997). Analytical toxicology: for clinical, forensic, and pharmaceutical chemists. Berlin: Walter de Gruyter. p. 243. ISBN 978-3-11-010731-9. Retrieved 2009-05-10.

- DDT and Its Derivatives: Environmental Aspects, Environmental Health Criteria monograph No. 83, Geneva: World Health Organization, ISBN 92-4-154283-7

- Michaels D (2008). Doubt is Their Product: How Industry's Assault on Science Threatens Your Health. New York: Oxford University Press. ISBN 978-0-19-530067-3.

- Walker C, Sibly RM, Hopkin S, Peakall DB (22 December 2005). Principles of ecotoxicology (3rd ed.). Boca Raton, FL: CRC/Taylor & Francis. pp. 300–. ISBN 978-0-8493-3635-5.

- Garrett, Laurie (31 October 1994). The Coming Plague: Newly Emerging Diseases in a World Out of Balance. Farrar, Straus and Giroux. p. 51. ISBN 978-1-4299-5327-6.

- Gutstein D (2009). Not a Conspiracy Theory: How Business Propaganda is Hijacking Democracy. ISBN 978-1-55470-191-9.. Relevant excerpt at Gutstein D (January 22, 2010). "Inside the DDT Propaganda Machine". The Tyee. Retrieved January 22, 2010.

- Francis Borgio J, Sahayaraj K and Alper Susurluk I (eds) . Microbial Insecticides: Principles and Applications, Nova Publishers, USA. 492pp. ISBN 978-1-61209-223-2

- Shaner, D. L.; Leonard, P. (2001). "Regulatory aspects of resistance management for herbicides and other crop protection products". In Powles, S. B.; Shaner, D. L. Herbicide Resistance and World Grains. CRC Press, Boca Raton, FL. pp. 279–294. ISBN 9781420039085.

- Moss, S. R. (2002). "Herbicide-Resistant Weeds". In Naylor,, R. E. L. Weed management handbook (9th ed.). Blackwell Science Ltd. pp. 225–252. ISBN 0-632-05732-7.

- "Endrin". Immediately Dangerous to Life or Health Concentrations (IDLH). National Institute for Occupational Safety and Health (NIOSH). 4 December 2014. Retrieved 19 March 2015.

- van Esch, G. T.; van Heemstra-Lequin, E. A. H. (1992). "Environmental Health Criteria 130: Endrin". International Programme on Chemical Safety. World Health Organization. Retrieved March 2015.

5

Organic Pollutants and their Effects

The text explains organic pollutants and their effects on our environment. Some of the topics covered are bioaccumulation, biomagnification, bioconcentration and environmental impacts of pesticides. The collection of pesticides and other chemicals in an organism is known as bioaccumulation whereas the effects that pesticides leave on species is discussed in the segment related to environmental impact of pesticides. The following chapter explains the importance of understanding organic pollutants and their effects on the environment.

Bioaccumulation

Bioaccumulation refers to the accumulation of substances, such as pesticides, or other chemicals in an organism. Bioaccumulation occurs when an organism absorbs a - possibly toxic - substance at a rate faster than that at which the substance is lost by catabolism and excretion. Thus, the longer the biological half-life of a toxic substance the greater the risk of chronic poisoning, even if environmental levels of the toxin are not very high. Bioaccumulation, for example in fish, can be predicted by models. Hypotheses for molecular size cutoff criteria for use as bioaccumulation potential indicators are not supported by data. Biotransformation can strongly modify bioaccumulation of chemicals in an organism.

Bioconcentration is a related but more specific term, referring to uptake and accumulation of a substance from water alone. By contrast, bioaccumulation refers to uptake from all sources combined (e.g. water, food, air, etc.)

Examples

An example of poisoning in the workplace can be seen from the phrase "as mad as a hatter" (18th and 19th century England). The process for stiffening the felt used in making hats more than a hundred years ago involved mercury, which forms organic species such as methylmercury, which is lipid-soluble, and tends to accumulate in the brain, resulting in mercury poisoning. Other lipid-soluble (fat-soluble) poisons include tetraethyllead compounds (the lead in leaded petrol), and DDT. These compounds are stored in the body's fat, and when the fatty tissues are used for energy, the compounds are released and cause acute poisoning.

Strontium-90, part of the fallout from atomic bombs, is chemically similar enough to calcium that it is utilized in osteogenesis, where its radiation can cause damage for a long time.

Naturally produced toxins can also bioaccumulate. The marine algal blooms known as "red tides" can result in local filter feeding organisms such as mussels and oysters becoming toxic; coral fish can be responsible for the poisoning known as ciguatera when they accumulate a toxin called ciguatoxin from reef algae.

Some animal species exhibit bioaccumulation as a mode of defense; by consuming toxic plants or animal prey, a species may accumulate the toxin, which then presents a deterrent to a potential predator. One example is the tobacco hornworm, which concentrates nicotine to a toxic level in its body as it consumes tobacco plants. Poisoning of small consumers can be passed along the food chain to affect the consumers later on. Other compounds that are not normally considered toxic can be accumulated to toxic levels in organisms. The classic example is of Vitamin A, which becomes concentrated in carnivore livers of e.g. polar bears: as a pure carnivore that feeds on other carnivores (seals), they accumulate extremely large amounts of Vitamin A in their livers. It was known by the native peoples of the Arctic that the livers of carnivores should not be eaten, but Arctic explorers have suffered Hypervitaminosis A from eating the bear livers (and there has been at least one example of similar poisoning of Antarctic explorers eating husky dog livers). One notable example of this is the expedition of Sir Douglas Mawson, where his exploration companion died from eating the liver of one of their dogs.

Coastal fish (such as the smooth toadfish) and seabirds (such as the Atlantic puffin) are often monitored for heavy metal bioaccumulation.

In some eutrophic aquatic systems, biodilution can occur. This trend is a decrease in a contaminant with an increase in trophic level and is due to higher concentrations of algae and bacteria to "dilute" the concentration of the pollutant.

Biomagnification

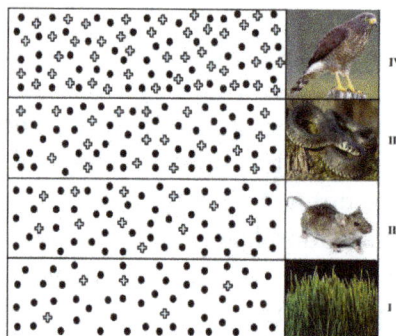

In biomagnification the concentration of the persistent toxins (crosses) increase higher up the food chain.

Biomagnification, also known as bioamplification or biological magnification, is the increasing concentration of a substance, such as a toxic chemical, in the tissues of organisms at successively higher levels in a food chain. This increase can occur as a result of:

- Persistence – where the substance can't be broken down by environmental processes

- Food chain energetics – where the substance concentration increases progressively as it moves up a food chain

- Low or non-existent rate of internal degradation or excretion of the substance – often due to water-insolubility

Biological magnification often refers to the process whereby certain substances such as pesticides or heavy metals move up the food chain, work their way into rivers or lakes, and are eaten by aquatic organisms such as fish, which in turn are eaten by large birds, animals or humans. The substances become concentrated in tissues or internal organs as they move up the chain. Bioaccumulants are substances that increase in concentration in living organisms as they take in contaminated air, water, or food because the substances are very slowly metabolized or excreted.

The following is an example showing how bio-magnification takes place in nature: An anchovy eats zoo-plankton that have tiny amounts of mercury that the zoo-plankton has picked up from the water throughout the anchovies lifespan. A tuna eats many of these anchovies over its life, accumulating the mercury in each of those anchovies into its body. If the mercury stunts the growth of the anchovies, that tuna is required to eat more little fish to stay alive. Because there are more little fish being eaten, the mercury content is magnified.

Processes

Although sometimes used interchangeably with "bioaccumulation", an important distinction is drawn between the two, and with bioconcentration.

- Bioaccumulation occurs *within* a trophic level, and is the increase in concentration of a substance in certain tissues of organisms' bodies due to absorption from food and the environment.

- Bioconcentration is defined as occurring when uptake from the water is greater than excretion.

Thus, bioconcentration and bioaccumulation occur within an organism, and biomagnification occurs across trophic (food chain) levels.

Biodilution is also a process that occurs to all trophic levels in an aquatic environment; it is the opposite of biomagnification, thus a pollutant gets smaller in concentration as it progresses up a food web.

Lipid, (lipophilic) or fat soluble substances cannot be diluted, broken down, or excreted in urine, a water-based medium, and so accumulate in fatty tissues of an organism if the organism lacks enzymes to degrade them. When eaten by another organism, fats are absorbed in the gut, carrying the substance, which then accumulates in the fats of the predator. Since at each level of the food chain there is a lot of energy loss, a predator must consume many prey, including all of their lipophilic substances.

For example, though mercury is only present in small amounts in seawater, it is absorbed by algae (generally as methylmercury). It is efficiently absorbed, but only very slowly excreted by organisms. Bioaccumulation and bioconcentration result in buildup in the adipose tissue of successive trophic levels: zooplankton, small nekton, larger fish, etc. Anything which eats these fish also consumes the higher level of mercury the fish have accumulated. This process explains why predatory fish such as swordfish and sharks or birds like osprey and eagles have higher concentrations of mercury in their tissue than could be accounted for by direct exposure alone. For example, herring contains mercury at approximately 0.01 parts per million (ppm) and shark contains mercury at greater than 1 ppm.

DDT is thought to biomagnify and biomagnification is one of the most significant reasons it was deemed harmful to the environment by the EPA and other organizations. DDT is stored in the fat of animals and takes many years to break down, and as the fat is consumed by predators, the amounts of DDT biomagnify. DDT is now a banned substance in many parts of the world.

Current Status

In a review of a large number of studies, Suedel et al. concluded that although biomagnification is probably more limited in occurrence than previously thought, there is good evidence that DDT, DDE, PCBs, toxaphene, and the organic forms of mercury and arsenic do biomagnify in nature. For other contaminants, bioconcentration and bioaccumulation account for their high concentrations in organism tissues. More recently, Gray reached a similar substances remaining in the organisms and not being diluted to non-threatening concentrations. The success of top predatory-bird recovery (bald eagles, peregrine falcons) in North America following the ban on DDT use in agriculture is testament to the importance of biomagnification.

Substances that Biomagnify

There are two main groups of substances that biomagnify. Both are lipophilic and not easily degraded. Novel organic substances are not easily degraded because organisms lack previous exposure and have thus not evolved specific detoxification and excretion mechanisms, as there has been no selection pressure from them. These substances are consequently known as "persistent organic pollutants" or POPs.

Metals are not degradable because they are elements. Organisms, particularly those subject to naturally high levels of exposure to metals, have mechanisms to sequester and excrete metals. Problems arise when organisms are exposed to higher concentrations than usual, which they cannot excrete rapidly enough to prevent damage. Some persistent heavy metals are especially dangerous and harmful to the organism's reproductive system.

Bioconcentration

Bioconcentration is the accumulation of a chemical in or on an organism when the source of chemical is solely water. Bioconcentration is a term that was created for use in the field of aquatic toxicology. Bioconcentration can also be defined as the process by which a chemical concentration in an aquatic organism exceeds that in water as a result of exposure to a waterborne chemical.

There are several ways in which to measure and assess bioaccumulation and bioconcentration. These include: octanol-water partition coefficients (K_{ow}), bioconcentration factors (BCF), bioaccumulation factors (BAF) and biota-sediment accumulation factor (BSAF). Each of these can be calculated using either empirical data or measurements as well as from mathematical models. One of these mathematical models is a fugacity-based BCF model developed by Don Mackay.

Bioconcentration factor can also be expressed as the ratio of the concentration of a chemical in an organism to the concentration of the chemical in the surrounding environment. The BCF is a measure of the extent of chemical sharing between an organism and the surrounding environment.

In surface water, the BCF is the ratio of a chemical's concentration in an organism to the chemical's aqueous concentration. BCF is often expressed in units of liter per kilogram (ratio of mg of chemical per kg of organism to mg of chemical per liter of water). BCF can simply be an observed ratio, or it can be the prediction of a partitioning model. A partitioning model is based on assumptions that chemicals partition between water and aquatic organisms as well as the idea that chemical equilibrium exists between the organisms and the aquatic environment in which it is found

Calculation

Bioconcentration can be described by a bioconcentration factor (BCF), which is the ratio of the chemical concentration in an organism or biota to the concentration in water:

$$BCF = \frac{Concentration_{Biota}}{Concentration_{Water}}$$

Bioconcentration factors can also be related to the octanol-water partition coefficient, K_{ow}. The octanol-water partition coefficient (K_{ow}) is correlated with the potential for a chemical to bioaccumulate in organisms; the BCF can be predicted from log K_{ow}, via computer programs based on structure activity relationship (SAR) or through the linear equation:

$$logBCF = mlogK_{OW} + b$$

Where:

$$K_{OW} = \frac{Concentration_{octanol}}{Concentration_{water}} = \frac{C_O}{C_W} \text{ at equilibrium}$$

Fugacity Capacity

Fugacity and BCF relate to each other in the following equation:

$$Z_{Fish} = \frac{P_{Fish} \times BCF}{H}$$

where Z_{Fish} is equal to the Fugacity capacity of a chemical in the fish, P_{Fish} is equal to the density of the fish (mass/length³), BCF is the partition coefficient between the fish and the water (length³/mass) and H is equal to the Henry's law constant (Length²/Time²)

Regression Equations for Estimations in Fish

Equation	Chemicals Used to obtain equation	Species Used
$logBCF = 0.76logKow - 0.23$	84	Fathead Minnow, Bluegill Sunfish, Rainbow Trout, Mosquitofish
$logBCF = logKow - 1.32$	44	Various
$logBCF = 2.791 - 0.564logS (S = watersolubility)$	36	Brook trout, Rainbow trout, Bluegill Sunfish, Fathead minnow, Carp
$logBCF = 3.41 - 0.508logS$	7	Various
$logBCF = 1.119logKoc - 1.579$	13	Various

Uses

Regulatory Uses

Through the use of the PBT Profiler and using criteria set forth by the United States Environmental Protection Agency under the Toxic Substances Control Act (TSCA), a substance is considered to be not bioaccumulative if it has a BCF less than 1000, bioaccumulative if it has a BCF from 1000–5000 and very bioaccumulative if it has a BCF greater than 5,000.

The thresholds under REACH are a BCF of > 2000 l/kg bzw. for the B and 5000 l/kg for vB criteria.

Applications

A bioconcentration factor greater than 1 is indicative of a hydrophobic or lipophilic chemical. It is an indicator of how probable a chemical is to bioaccumulate. These chemicals have high lipid affinities and will concentrate in tissues with high lipid content instead of in an aqueous environment like the cytosol. Models are used to predict chemical partitioning in the environment which in turn allows the prediction of the biological fate of lipophilic chemicals.

Equilibrium Partitioning Models

Based on an assumed steady state scenario, the fate of a chemical in a system is modeled giving predicted endpoint phases and concentrations.

It needs to be considered that reaching steady state may need a substantial amount of time as estimated using the following equation (in hours).

$$t_{eSS} = 0.00654 \cdot K_{OW} + 55.31$$

For a substance with a K_{OW} of 4, it thus takes approximately five days to reach effective steady state. For a K_{OW} of 6, the equilibrium time increases to nine months.

Fugacity Models

Fugacity is another predictive criterion for equilibrium among phases that has units of pressure. It is equivalent to partial pressure for most environmental purposes. It is the absconding propensity of a material. BCF can be determined from output parameters of a fugacity model and thus used to predict the fraction of chemical immediately interacting with and possibly having an effect on an organism.

Food Web Models

If organism-specific fugacity values are available, it is possible to create a food web model which takes trophic webs into consideration. This is especially pertinent for conservative chemicals that are not easily metabolized into degradation products. Biomagnification of conservative chemicals such as toxic metals can be harmful to apex predators like orca whales, osprey, and bald eagles.

Applications to Toxicology

Predictions

Bioconcentration factors facilitate predicting contamination levels in an organism based on chemical concentration in surrounding water. BCF in this setting only applies to aquatic organisms. Air breathing organisms do not take up chemicals in the same manner as other aquatic organisms. Fish, for example uptake chemicals via ingestion and osmotic gradients in gill lamellae.

When working with benthic macroinvertebrates, both water and benthic sediments may contain chemical that affects the organism. Biota-sediment accumulation factor (BSAF) and biomagnification factor (BMF) also influence toxicity in aquatic environments.

BCF does not explicitly take metabolism into consideration so it needs to be added to models at other points through uptake, elimination or degradation equations for a selected organism.

Body Burden

Chemicals with high BCF values are more lipophilic, and at equilibrium organisms will have greater concentrations of chemical than other phases in the system. Body burden is the total amount of chemical in the body of an organism, and body burdens will be greater when dealing with a lipophilic chemical.

Biological Factors

In determining the degree at which bioconcentration occurs biological factors have to be kept in mind.The rate at which an organism is exposed through respiratory surfaces and contact with dermal surfaces of the organism, competes against the rate of excretion from an organism. The rate of excretion is a loss of chemical from the respiratory surface, growth dilution, fecal excretion, and metabolic biotransformation. Growth dilution is not an actual process of excretion but due to the mass of the organism increasing while the contaminant concentration remains constant dilution occurs.

The interaction between inputs and outputs is shown here:

$$\frac{dC_B}{dt} = (k_1 C_{WD}) - (k_2 + k_E + k_M + k_G)C_B$$

The variables are defined as:$C_B$is the concentration in the organism ($g*kg^{-1}$). t represents a unit of time (d^{-1}). k_1 is the rate constant for chemical uptake from water at the respiratory surface (L*kg^{-1}*d^{-1}). C$_{WD}$ is the chemical concentration dissolved in water (g*L^{-1}). k$_2$,k$_E$,k$_G$,k$_B$ are rate constants that represent excretion from the organism from the respiratory surface, fecal excretion, metabolic transformation, and growth dilution (d^{-1}).

Static variables influence BCF as well. Because organisms are modeled as bags of fat, lipid to water ratio is a factor that needs to be considered. Size also plays a role as the surface to volume ratio influence the rate of uptake from the surrounding water. The species of concern is a primary factor in influencing BCF values due to it determining all of the biological factors that alter a BCF.

Environmental Parameters

Temperature

Temperature may affect metabolic transformation, and bioenergetics. An example of this is the movement of the organism may change as well as rates of excretion. If a contaminant is ionic, the change in pH that is influenced by a change in temperature may also influence the bioavailability

Water Quality

The natural particle content as well as organic carbon content in water can affect the bioavailability. The contaminant can bind to the particles in the water, making uptake more difficult, as well as become ingested by the organism. This ingestion could consist of contaminated particles which would cause the source of contamination to be from more than just water.

Environmental Impact of Pesticides

Preparing to spray a hazardous pesticide

The environmental impact of pesticides consists of the effects of pesticides on non-target species. Over 98% of sprayed insecticides and 95% of herbicides reach a destination other than their target species, because they are sprayed or spread across entire agricultural fields. Runoff can carry pesticides into aquatic environments while wind can carry them to other fields, grazing areas, human settlements and undeveloped areas, potentially affecting other species. Other problems emerge from poor production, transport and storage practices. Over time, repeated application increases pest resistance, while its effects on other species can facilitate the pest's resurgence.

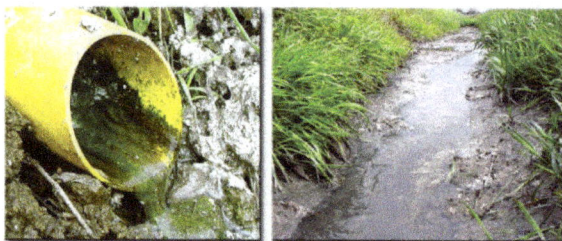

Drainage of fertilizers and pesticides into a stream

Each pesticide or pesticide class comes with a specific set of environmental concerns. Such undesirable effects have led many pesticides to be banned, while regulations have limited and/or reduced the use of others. Over time, pesticides have generally become less persistent and more species-specific, reducing their environmental footprint. In addition the amounts of pesticides applied per hectare have declined, in some cases by 99%. However, the global spread of pesticide use, including the use of older/obsolete pesticides that have been banned in some jurisdictions, has increased overall.

Agriculture and the Environment

The arrival of humans in an area, to live or to conduct agriculture, necessarily has environmental impacts. These range from simple crowding out of wild plants in favor of more desirable cultivars to larger scale impacts such as reducing biodiversity by reducing food availability of native species, which can propagate across food chains. The use of agricultural chemicals such as fertilizer and pesticides magnify those impacts. While advances in agrochemistry have reduced those impacts, for example by the replacement of long-lived chemicals with those that reliably degrade, even in the best case they remain substantial. These effects are magnified by the use of older chemistries and poor management practices.

History

While concern ecotoxicology began with acute poisoning events in the late 19th century; public concern over the undesirable environmental effects of chemicals arose in the early 1960s with the publication of Rachel Carson's book, *Silent Spring*. Shortly thereafter, DDT, originally used to combat malaria, and its metabolites were shown to cause population-level effects in raptorial birds. Initial studies in industrialized countries focused on acute mortality effects mostly involving birds or fish.

Data on pesticide usage remain scattered and/or not publicly available (3). The common practice of incident registration is inadequate for understanding the entirety of effects.

Since 1990, research interest has shifted from documenting incidents and quantifying chemical exposure to studies aimed at linking laboratory, mesocosm and field experiments. The proportion of effect-related publications has increased. Animal studies mostly focus on fish, insects, birds, amphibians and arachnids.

Since 1993, the United States and the European Union have updated pesticide risk assessments, ending the use of acutely toxic organophosphate and carbamate insecticides. Newer pesticides aim at efficiency in target and minimum side effects in nontarget organisms. The phylogenetic proximity of beneficial and pest species complicates the project.

One of the major challenges is to link the results from cellular studies through many levels of increasing complexity to ecosystems.

Specific Pesticide Effects

Pesticide environmental effects	
Pesticide/class	**Effect(s)**
Organochlorine DDT/DDE	Egg shell thinning in raptorial birds
	Endocrine disruptor
	Thyroid disruption properties in rodents, birds, amphibians and fish
	Acute mortality attributed to inhibition of acetylcholine esterase activity
:DDT	Carcinogen
	Endocrine disruptor
:DDT/Diclofol, Dieldrin and Toxaphene	Juvenile population decline and adult mortality in wildlife reptiles
:DDT/Toxaphene/Parathion	Susceptibility to fungal infection
Triazine	Earthworms became infected with monocystid gregarines
:Chlordane	Interact with vertebrate immune systems

Carbamates, the phenoxy herbicide 2,4-D, and atrazine	Interact with vertebrate immune systems
Anticholinesterase	Bird poisoning
	Animal infections, disease outbreaks and higher mortality.
Organophosphate	Thyroid disruption properties in rodents, birds, amphibians and fish
	Acute mortality attributed to inhibition of acetylcholine esterase activity
	Immunotoxicity, primarily caused by the inhibition of serine hydrolases or esterases
	Oxidative damage
	Modulation of signal transduction pathways
	Impaired metabolic functions such as thermoregulation, water and/or food intake and behavior, impaired development, reduced reproduction and hatching success in vertebrates.
Carbamate	Thyroid disruption properties in rodents, birds, amphibians and fish
	Impaired metabolic functions such as thermoregulation, water and/or food intake and behavior, impaired development, reduced reproduction and hatching success in vertebrates.
	Interact with vertebrate immune systems
	Acute mortality attributed to inhibition of acetylcholine esterase activity
Phenoxy herbicide 2,4-D	Interact with vertebrate immune systems
Atrazine	Interact with vertebrate immune systems
	Reduced northern leopard frog (Rana pipiens) populations because atrazine killed phytoplankton, thus allowing light to penetrate the water column and periphyton to assimilate nutrients released from the plankton. Periphyton growth provided more food to grazers, increasing snail populations, which provide intermediate hosts for trematode.
Pyrethroid	Thyroid disruption properties in rodents, birds, amphibians and fish
Thiocarbamate	Thyroid disruption properties in rodents, birds, amphibians and fish
Triazine	Thyroid disruption properties in rodents, birds, amphibians and fish

Triazole	Thyroid disruption properties in rodents, birds, amphibians and fish
	Impaired metabolic functions such as thermoregulation, water and/or food intake and behavior, impaired development, reduced reproduction and hatching success in vertebrates.
Nicotinoid	respiratory, cardiovascular, neurological, and immunological toxicity in rats and humans
	Disrupt biogenic amine signaling and cause subsequent olfactory dysfunction, as well as affecting foraging behavior, learning and memory.
Imidacloprid, Imidacloprid/pyrethroid λ-cyhalothrin	Impaired foraging, brood development, and colony success in terms of growth rate and new queen production.
Thiamethoxa	High honey bee worker mortality due to homing failure (risks for colony collapse remain controversial)
Spinosyns	Affect various physiological and behavioral traits of beneficial arthropods, particularly hymenopterans
Bt corn/Cry	Reduced abundance of some insect taxa, predominantly susceptible Lepidopteran herbivores as well as their predators and parasitoids.
Herbicide	Reduced food availability and adverse secondary effects on soil invertebrates and butterflies
	Decreased species abundance and diversity in small mammals.
Benomyl	Altered the patch-level floral display and later a two-thirds reduction of the total number of bee visits and in a shift in the visitors from large-bodied bees to small-bodied bees and flies
Herbicide and planting cycles	Reduced survival and reproductive rates in seed-eating or carnivorous birds

Air

Spraying a mosquito pesticide over a city

Pesticides can contribute to air pollution. Pesticide drift occurs when pesticides suspended in the air as particles are carried by wind to other areas, potentially contaminating them. Pesticides that are applied to crops can volatilize and may be blown by winds into nearby areas, potentially posing a threat to wildlife. Weather conditions at the time of application as well as temperature and relative humidity change the spread of the pesticide in the air.

As wind velocity increases so does the spray drift and exposure. Low relative humidity and high temperature result in more spray evaporating. The amount of inhalable pesticides in the outdoor environment is therefore often dependent on the season. Also, droplets of sprayed pesticides or particles from pesticides applied as dusts may travel on the wind to other areas, or pesticides may adhere to particles that blow in the wind, such as dust particles. Ground spraying produces less pesticide drift than aerial spraying does. Farmers can employ a buffer zone around their crop, consisting of empty land or non-crop plants such as evergreen trees to serve as windbreaks and absorb the pesticides, preventing drift into other areas. Such windbreaks are legally required in the Netherlands.

Pesticides that are sprayed on to fields and used to fumigate soil can give off chemicals called volatile organic compounds, which can react with other chemicals and form a pollutant called tropospheric ozone. Pesticide use accounts for about 6 percent of total tropospheric ozone levels.

Water

In the United States, pesticides were found to pollute every stream and over 90% of wells sampled in a study by the US Geological Survey. Pesticide residues have also been found in rain and groundwater. Studies by the UK government showed that pesticide concentrations exceeded those allowable for drinking water in some samples of river water and groundwater.

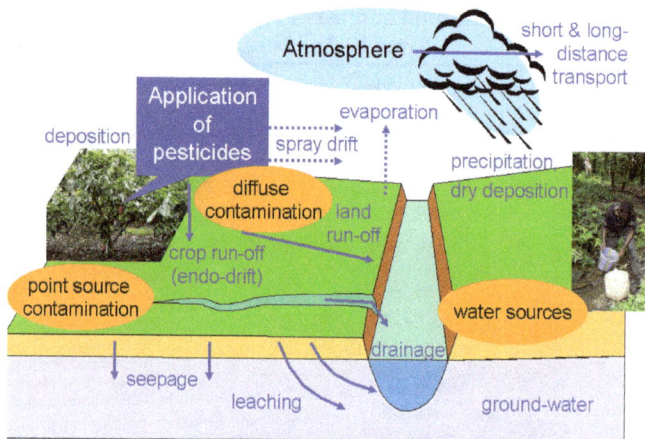

Pesticide pathways

Pesticide impacts on aquatic systems are often studied using a hydrology transport model to study movement and fate of chemicals in rivers and streams. As early as the 1970s quantitative analysis of pesticide runoff was conducted in order to predict amounts of pesticide that would reach surface waters.

There are four major routes through which pesticides reach the water: it may drift outside of the intended area when it is sprayed, it may percolate, or leach, through the soil, it may be carried to the water as runoff, or it may be spilled, for example accidentally or through neglect. They may also be carried to water by eroding soil. Factors that affect a pesticide's ability to contaminate water include its water solubility, the distance from an application site to a body of water, weather, soil type, presence of a growing crop, and the method used to apply the chemical.

Maximum limits of allowable concentrations for individual pesticides in public bodies of water are set by the Environmental Protection Agency in the US. Similarly, the government of the United Kingdom sets Environmental Quality Standards (EQS), or maximum allowable concentrations of some pesticides in bodies of water above which toxicity may occur. The European Union also regulates maximum concentrations of pesticides in water.

Soil

Many of the chemicals used in pesticides are persistent soil contaminants, whose impact may endure for decades and adversely affect soil conservation.

Caution against entering a field sprayed with sulphuric acid

The use of pesticides decreases the general biodiversity in the soil. Not using the chemicals results in higher soil quality, with the additional effect that more organic matter in the soil allows for higher water retention. This helps increase yields for farms in drought years, when organic farms have had yields 20-40% higher than their conventional counterparts. A smaller content of organic matter in the soil increases the amount of pesticide that will leave the area of application, because organic matter binds to and helps break down pesticides.

Degradation and sorption are both factors which influence the persistence of pesticides in soil. Depending on the chemical nature of the pesticide, such processes control directly the transportation from soil to water, and in turn to air and our food. Breaking down organic substances, degradation, involves interactions among microorganisms in the soil. Sorption affects bioaccumulation of pesticides which are dependent on organic matter in the soil. Weak organic acids have been shown to be weakly sorbed by soil, because of pH and mostly acidic structure. Sorbed chemicals have been shown to be less accessible to microorganisms. Aging mechanisms are poorly understood but as residence times in soil increase, pesticide residues become more resistant to degradation and extraction as they lose biological activity.

Effect on Plants

Nitrogen fixation, which is required for the growth of higher plants, is hindered by pesticides in soil. The insecticides DDT, methyl parathion, and especially pentachlorophenol have been shown to interfere with legume-rhizobium chemical signaling. Reduction of this symbiotic chemical signaling results in reduced nitrogen fixation and thus reduced crop yields. Root nodule formation in these plants saves the world economy $10 billion in synthetic nitrogen fertilizer every year.

Crop spraying

Pesticides can kill bees and are strongly implicated in pollinator decline, the loss of species that pollinate plants, including through the mechanism of Colony Collapse Disorder, in which worker bees from a beehive or western honey bee colony abruptly disappear. Application of pesticides to crops that are in bloom can kill honeybees, which act as pollinators. The USDA and USFWS estimate that US farmers lose at least $200 million a year from reduced crop pollination because pesticides applied to fields eliminate about a fifth of honeybee colonies in the US and harm an additional 15%.

On the other side, pesticides have some direct harmful effect on plant including poor root hair development, shoot yellowing and reduced plant growth.

Effect on Animals

Many kinds of animals are harmed by pesticides, leading many countries to regulate pesticide usage through Biodiversity Action Plans.

Animals including humans may be poisoned by pesticide residues that remain on food, for example when wild animals enter sprayed fields or nearby areas shortly after spraying.

Pesticides can eliminate some animals' essential food sources, causing the animals to relocate, change their diet or starve. Residues can travel up the food chain; for example, birds can be harmed when they eat insects and worms that have consumed pesticides. Earthworms digest organic matter and increase nutrient content in the top layer of soil. They protect human health by ingesting decomposing litter and serving as bioindicators of soil activity. Pesticides have had harmful effects on growth and reproduction on earthworms. Some pesticides can bioaccumulate, or build up to toxic levels in the bodies of organisms that consume them over time, a phenomenon that impacts species high on the food chain especially hard.

Birds

The US Fish and Wildlife Service estimates that 72 million birds are killed by pesticides in the United States each year. Bald eagles are common examples of nontarget organisms that are impacted by pesticide use. Rachel Carson's book *Silent Spring* dealt with damage to bird species due to pesticide bioaccumulation. There is evidence that birds are continuing to be harmed by pesticide use. In the farmland of the United Kingdom, populations of ten different bird species declined by 10 million breeding individuals between 1979 and 1999, allegedly from loss of plant and invertebrate species on which the birds feed. Throughout Europe, 116 species of birds were threatened as of 1999. Reductions in bird populations have been found to be associated with times and areas in which pesticides are used. DDE-induced egg shell thinning has especially affected European and North American bird populations. In another example, some types of fungicides used in peanut farming are only slightly toxic to birds and mammals, but may kill earthworms, which can in turn reduce populations of the birds and mammals that feed on them.

In England, the use of pesticides in gardens and farmland has seen a reduction in the number of common chaffinches

Some pesticides come in granular form. Wildlife may eat the granules, mistaking them for grains of food. A few granules of a pesticide may be enough to kill a small bird.

The herbicide paraquat, when sprayed onto bird eggs, causes growth abnormalities in embryos and reduces the number of chicks that hatch successfully, but most herbicides do not directly cause much harm to birds. Herbicides may endanger bird populations by reducing their habitat.

Aquatic Life

Fish and other aquatic biota may be harmed by pesticide-contaminated water. Pesticide surface runoff into rivers and streams can be highly lethal to aquatic life, sometimes killing all the fish in a particular stream.

Using an aquatic herbicide

Application of herbicides to bodies of water can cause fish kills when the dead plants decay and consume the water's oxygen, suffocating the fish. Herbicides such as copper sulfite that are applied to water to kill plants are toxic to fish and other water animals at concentrations similar to those used to kill the plants. Repeated exposure to sublethal doses of some pesticides can cause physiological and behavioral changes that reduce fish populations, such as abandonment of nests and broods, decreased immunity to disease and decreased predator avoidance.

Wide field margins can reduce fertilizer and pesticide pollution in streams and rivers

Application of herbicides to bodies of water can kill plants on which fish depend for their habitat.

Pesticides can accumulate in bodies of water to levels that kill off zooplankton, the main source of food for young fish. Pesticides can also kill off insects on which some fish feed, causing the fish to travel farther in search of food and exposing them to greater risk from predators.

The faster a given pesticide breaks down in the environment, the less threat it poses to aquatic life. Insecticides are typically more toxic to aquatic life than herbicides and fungicides.

Amphibians

In the past several decades, amphibian populations have declined across the world, for unexplained reasons which are thought to be varied but of which pesticides may be a part.

Pesticide mixtures appear to have a cumulative toxic effect on frogs. Tadpoles from ponds containing multiple pesticides take longer to metamorphose and are smaller when they do, decreasing their ability to catch prey and avoid predators. Exposing tadpoles to the organochloride endosulfan at levels likely to be found in habitats near fields sprayed with the chemical kills the tadpoles and causes behavioral and growth abnormalities.

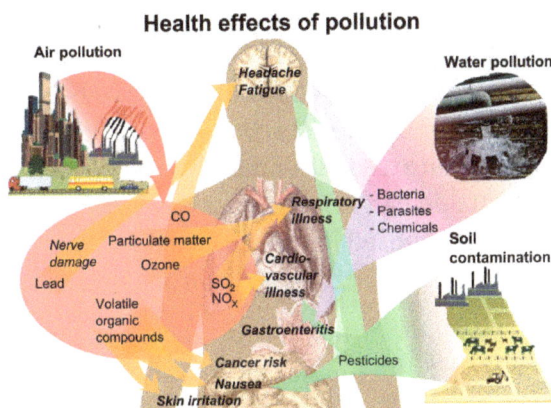

Pesticides are implicated in a range of impacts on human health due to pollution

The herbicide atrazine can turn male frogs into hermaphrodites, decreasing their ability to reproduce. Both reproductive and nonreproductive effects in aquatic reptiles and amphibians have been reported. Crocodiles, many turtle species and some lizards lack sex-distinct chromosomes until after fertilization during organogenesis, depending on temperature. Embryonic exposure in turtles to various PCBs causes a sex reversal. Across the United States and Canada disorders such as decreased hatching success, feminization, skin lesions, and other developmental abnormalities have been reported.

Humans

Pesticides can enter the body through inhalation of aerosols, dust and vapor that contain pesticides; through oral exposure by consuming food/water; and through skin exposure by direct contact. Pesticides secrete into soils and groundwater which can end up in drinking water, and pesticide spray can drift and pollute the air.

The effects of pesticides on human health depend on the toxicity of the chemical and the length and magnitude of exposure. Farm workers and their families experience the greatest exposure to agricultural pesticides through direct contact. Every human contains pesticides in their fat cells.

Children are more susceptible and sensitive to pesticides, because they are still developing and have a weaker immune system than adults. Children may be more exposed due to their closer proximity to the ground and tendency to put unfamiliar objects in their mouth. Hand to mouth contact depends on the child's age, much like lead exposure. Children under the age of six months are more apt to experience exposure from breast milk and inhalation of small particles. Pesticides tracked into the home from family members increase the risk of exposure. Toxic residue in food may contribute to a child's exposure. The chemicals can bioaccumulate in the body over time.

Exposure effects can range from mild skin irritation to birth defects, tumors, genetic changes, blood and nerve disorders, endocrine disruption, coma or death. Developmental effects have been associated with pesticides. Recent increases in childhood cancers in throughout North America, such as leukemia, may be a result of somatic cell mutations. Insecticides targeted to disrupt insects can have harmful effects on mammalian nervous systems. Both chronic and acute alterations have been observed in exposees. DDT and its breakdown product DDE disturb estrogenic activity and possibly lead to breast cancer. Fetal DDT exposure reduces male penis size in animals and can produce undescended testicles. Pesticide can affect fetuses in early stages of development, in utero and even if a parent was exposed before conception. Reproductive disruption has the potential to occur by chemical reactivity and through structural changes.

Persistent Organic Pollutants

Persistent organic pollutants (POPs) are compounds that resist degradation and thus remain in the environment for years. Some pesticides, including aldrin, chlordane, DDT, dieldrin, endrin, heptachlor, hexachlorobenzene, mirex and toxaphene, are considered POPs. Some POPs have the ability to volatilize and travel great distances through the atmosphere to become deposited in remote regions. Such chemicals may have the ability to bioaccumulate and biomagnify and can bioconcentrate (i.e. become more concentrated) up to 70,000 times their original concentrations. POPs can affect non-target organisms in the environment and increase risk to humans by disruption in the endocrine, reproductive, and immune systems.

Pest Resistance

Pests may evolve to become resistant to pesticides. Many pests will initially be very susceptible to pesticides, but following mutations in their genetic makeup become resistant and survive to reproduce.

Resistance is commonly managed through pesticide rotation, which involves alternating among pesticide classes with different modes of action to delay the onset of or mitigate existing pest resistance.

Pest Rebound and Secondary Pest Outbreaks

Non-target organisms can also be impacted by pesticides. In some cases, a pest insect that is controlled by a beneficial predator or parasite can flourish should an insecticide application kill both pest and beneficial populations. A study comparing biological pest control and pyrethroid insecticide for diamondback moths, a major cabbage family insect pest, showed that the pest population rebounded due to loss of insect predators, whereas the biocontrol did not show the same effect. Likewise, pesticides sprayed to control mosquitoes may temporarily depress mosquito populations, however they may result in a larger population in the long run by damaging natural controls. This phenomenon, wherein the population of a pest species rebounds to equal or greater numbers than it had before pesticide use, is called pest resurgence and can be linked to elimination of its predators and other natural enemies.

Loss of predator species can also lead to a related phenomenon called secondary pest outbreaks, an increase in problems from species that were not originally a problem due to loss of their predators or parasites. An estimated third of the 300 most damaging insects in the US were originally secondary pests and only became a major problem after the use of pesticides. In both pest resurgence and secondary outbreaks, their natural enemies were more susceptible to the pesticides than the pests themselves, in some cases causing the pest population to be higher than it was before the use of pesticide.

Eliminating Pesticides

Many alternatives are available to reduce the effects pesticides have on the environment. Alternatives include manual removal, applying heat, covering weeds with plastic, placing traps and lures, removing pest breeding sites, maintaining healthy soils that breed healthy, more resistant plants, cropping native species that are naturally more resistant to native pests and supporting biocontrol agents such as birds and other pest predators. In the United States, conventional pesticide use peaked in 1979, and by 2007, had been reduced by 25 percent from the 1979 peak level, while US agricultural output increased by 43 percent over the same period.

Biological controls such as resistant plant varieties and the use of pheromones, have been successful and at times permanently resolve a pest problem. Integrated Pest Management (IPM) employs chemical use only when other alternatives are ineffective. IPM causes less harm to humans and the environment. The focus is broader than on a specific pest, considering a range of pest control alternatives. Biotechnology can also be an innovative way to control pests. Strains can be genetically modified (GM) to increase their resistance to pests. However the same techniques can be used to increase pesticide resistance and was employed by Monsanto to create glyphosate-resistant strains of major crops. In 2010, 70% of all the corn that was planted was resistant to glyphosate; 78% of cotton, and 93% of all soybeans.

References

- Landis WG, Sofield RM, Yu MH (2011). Introduction to Environmental Toxicology: Molecular Structures to Ecological Landscapes (Fourth ed.). Boca Raton, FL: CRC Press. pp. 117–162. ISBN 978-1-4398-0410-0.

- George Tyler Miller (1 January 2004). Sustaining the Earth: An Integrated Approach. Thomson/Brooks/Cole. pp. 211–216. ISBN 978-0-534-40088-0.

- Howell V. Daly; John T. Doyen; Alexander H. Purcell (1 January 1998). Introduction to Insect Biology and Diversity. Oxford University Press. pp. 279–300. ISBN 978-0-19-510033-4.

- Lorenz, Eric S. (2009). "Potential Health Effects of Pesticides" (PDF). Ag Communications and Marketing: 1–8. Retrieved February 2014.

- EPA. "Category for Persistent, Bioaccumulative, and Toxic New Chemical Substances". Federal Register Environmental Documents. USEPA. Retrieved 3 June 2012.

- Arnot, Jon; Frank A.P.C. Gobas (13 December 2006). "A review of bioconcentration factor (BCF) and bioaccumulation factor (BAF) assessments for organic chemicals in aquatic organisms" (PDF). NRC Canada. 12: 257–297. doi:10.1139/A06-005. Retrieved 7 June 2012.

6

Adverse Impacts of Organic Pollutants on Humans

Organic pollutants have adverse effects on humans as well as our environment. Endocrine disruptor and neurotoxins are toxins of organic pollutants; endocrine disruptors cause tumors and birth defects whereas neurotoxins are known for causing nervous system related problems. This chapter helps the reader in developing an in-depth understanding of the impact that organic pollutants have on humans.

Endocrine Disruptor

Endocrine disruptors are chemicals that, at certain doses, can interfere with endocrine (or hormone) systems. These disruptions can cause cancerous tumors, birth defects, and other developmental disorders. Any system in the body controlled by hormones can be derailed by hormone disruptors. Specifically, endocrine disruptors may be associated with the development of learning disabilities, severe attention deficit disorder, cognitive and brain development problems; deformations of the body (including limbs); breast cancer, prostate cancer, thyroid and other cancers; sexual development problems such as feminizing of males or masculinizing effects on females, etc. Recently The Endocrine Society released a statement on Endocrine-Disrupting Chemicals (EDCs) specifically listing obesity, diabetes, female reproduction, male reproduction, hormone-sensitive cancers in females, prostate cancer in males, thyroid, and neurodevelopment and neuroendocrine systems as being affected biological aspects of being exposed to EDCs. The critical period of development for most organisms is between the transition from a fertilized egg into a fully formed infant. As the cells begin to grow and differentiate, there are critical balances of hormones and protein changes that must occur. Therefore, a dose of disrupting chemicals may do substantial damage to a developing fetus. The same dose may not significantly affect adult mothers.

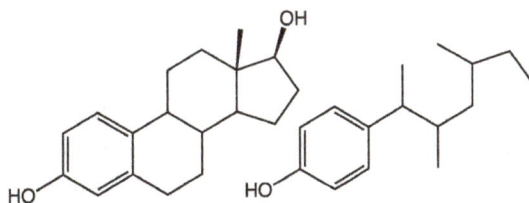

A comparison of the structures of the natural hormone estradiol (left) and one of the nonyl-phenols (right), an endocrine disruptor

There has been controversy over endocrine disruptors, with some groups calling for swift action by regulators to remove them from the market, and regulators and other scientists calling for further study. Some endocrine disruptors have been identified and removed from the market (for example, a drug called diethylstilbestrol), but it is uncertain whether some endocrine disruptors on the market actually harm humans and wildlife at the doses to which wildlife and humans are exposed. Additionally, a key scientific paper, published in the journal *Science*, which helped launch the movement of those opposed to endocrine disruptors, was retracted and its author found to have committed scientific misconduct.

Found in many household and industrial products, endocrine disruptors are substances that "interfere with the synthesis, secretion, transport, binding, action, or elimination of natural hormones in the body that are responsible for development, behavior, fertility, and maintenance of homeostasis (normal cell metabolism)." They are sometimes also referred to as hormonally active agents, endocrine disrupting chemicals, or endocrine disrupting compounds.

Studies in cells and laboratory animals have shown that EDCs can cause adverse biological effects in animals, and low-level exposures may also cause similar effects in human beings. The term *endocrine disruptor* is often used as synonym for xenohormone although the latter can mean any naturally occurring or artificially produced compound showing hormone-like properties (usually binding to certain hormonal receptors). EDCs in the environment may also be related to reproductive and infertility problems in wildlife and bans and restrictions on their use has been associated with a reduction in health problems and the recovery of some wildlife populations.

History

The term *endocrine disruptor* was coined at the Wingspread Conference Centre in Wisconsin, in 1991. One of the early papers on the phenomenon was by Theo Colborn in 1993. In this paper, she stated that environmental chemicals disrupt the development of the endocrine system, and that effects of exposure during development are often permanent. Although the endocrine disruption has been disputed by some, work sessions from 1992 to 1999 have generated consensus statements from scientists regarding the hazard from endocrine disruptors, particularly in wildlife and also in humans. The Endocrine Society released a scientific statement outlining mechanisms and effects of endocrine disruptors on "male and female reproduction, breast development and cancer, prostate cancer, neuroendocrinology, thyroid, metabolism and obesity, and cardiovascular endocrinology," and showing how experimental and epidemiological studies converge with human clinical observations "to implicate EDCs as a significant concern to public health." The statement noted that it is difficult to show that endocrine disruptors cause human diseases, and it recommended that the precautionary principle should be followed. A concurrent statement expresses policy concerns.

Endocrine disrupting compounds encompass a variety of chemical classes, including drugs, pesticides, compounds used in the plastics industry and in consumer products, industrial by-products and pollutants, and even some naturally produced botanical chemicals. Some are pervasive and widely dispersed in the environment and may bio-accumulate. Some are persistent organic pollutants (POP's), and can be transported long distances across national boundaries and have been found in virtually all regions of the world, and may even concentrate near the North Pole, due to weather patterns and cold conditions. Others are rapidly degraded in the environment or human body or may be present for only short periods of time. Health effects attributed to endocrine disrupting compounds include a range of reproductive problems (reduced fertility, male and female reproductive tract abnormalities, and skewed male/female sex ratios, loss of fetus, menstrual problems); changes in hormone levels; early puberty; brain and behavior problems; impaired immune functions; and various cancers.

One example of the consequences of the exposure of developing animals, including humans, to hormonally active agents is the case of the drug diethylstilbestrol (DES), a non-steroidal estrogen and not an environmental pollutant. Prior to its ban in the early 1970s, doctors prescribed DES to as many as five million pregnant women to block spontaneous abortion, an off-label use of this medication prior to 1947. It was discovered after the children went through puberty that DES affected the development of the reproductive system and caused vaginal cancer. The relevance of the DES saga to the risks of exposure to endocrine disruptors is questionable, as the doses involved are much higher in these individuals than in those due to environmental exposures.

Aquatic life subjected to endocrine disruptors in an urban effluent have experienced decreased levels of serotonin and increased feminization.

In 2013 the WHO and the United Nations Environment Programme released a study, the most comprehensive report on EDCs to date, calling for more research to fully understand the associations between EDCs and the risks to health of human and animal life. The team pointed to wide gaps in knowledge and called for more research to obtain a fuller picture of the health and environmental impacts of endocrine disruptors. To improve global knowledge the team has recommended:

- *Testing: known EDCs are only the 'tip of the iceberg' and more comprehensive testing methods are required to identify other possible endocrine disruptors, their sources, and routes of exposure.*

- *Research: more scientific evidence is needed to identify the effects of mixtures of EDCs on humans and wildlife (mainly from industrial by-products) to which humans and wildlife are increasingly exposed.*

- *Reporting: many sources of EDCs are not known because of insufficient reporting and information on chemicals in products, materials and goods.*

- *Collaboration: more data sharing between scientists and between countries can fill gaps in data, primarily in developing countries and emerging economies.*

Endocrine System

Endocrine systems are found in most varieties of animals. The endocrine system consists of glands that secrete hormones, and receptors that detect and react to the hormones.

Hormones travel throughout the body and act as chemical messengers. Hormones interface with cells that contain matching receptors in or on their surfaces. The hormone binds with the receptor, much like a key would fit into a lock. The endocrine system regulates adjustments through slower internal processes, using hormones as messengers. The endocrine system secretes hormones in response to environmental stimuli and to orchestrate developmental and reproductive changes. The adjustments brought on by the endocrine system are biochemical, changing the cell's internal and external chemistry to bring about a long term change in the body. These systems work together to maintain the proper functioning of the body through its entire life cycle. Sex steroids such as estrogens and androgens, as well as thyroid hormones, are subject to feedback regulation, which tends to limit the sensitivity of these glands.

Hormones work at very small doses (part per billion ranges). Endocrine disruption can thereby also occur from low-dose exposure to exogenous hormones or hormonally active chemicals that can interfere with receptors for other hormonally mediated processes. Furthermore, since endogenous hormones are already present in the body in biologically active concentrations, additional exposure to relatively small amounts of exogenous hormonally active substances can disrupt the proper functioning of the body's endocrine system. Thus, an endocrine disruptor can elicit adverse effects at much lower doses than a toxicity, acting through a different mechanism.

The timing of exposure is also critical. Most critical stages of development occur in utero, where the fertilized egg divides, rapidly developing every structure of a fully formed baby, including much of the wiring in the brain. Interfering with the hormonal communication in utero can have profound effects both structurally and toward brain development. Depending on the stage of reproductive development, interference with hormonal signaling can result in irreversible effects not seen in adults exposed to the same dose for the same length of time. Experiments with animals have identified critical developmental time points in utero and days after birth when exposure to chemicals that interfere with or mimic hormones have adverse effects that persist into adulthood. Disruption of thyroid function early in development may be the cause of abnormal sexual development in both males and females early motor development impairment, and learning disabilities.

There are studies of cell cultures, laboratory animals, wildlife, and accidentally exposed humans that show that environmental chemicals cause a wide range of reproductive, developmental, growth, and behavior effects, and so while "endocrine disruption in humans by pollutant chemicals remains largely undemonstrated, the underlying science is sound and the potential for such effects is real." While compounds that produce estrogenic, androgenic, antiandrogenic, and antithyroid actions have been studied, less is known about interactions with other hormones.

The interrelationship between exposures to chemicals and health effects are rather complex. It is hard to definitively link a particular chemical with a specific health effect, and exposed adults may not show any ill effects. But, fetuses and embryos, whose growth and development are highly controlled by the endocrine system, are more vulnerable to exposure and may suffer overt or subtle lifelong health and/or reproductive abnormalities. Prebirth exposure, in some cases, can lead to permanent alterations and adult diseases.

Some in the scientific community are concerned that exposure to endocrine disruptors in the womb or early in life may be associated with neurodevelopmental disorders including reduced IQ, ADHD, and autism. Certain cancers and uterine abnormalities in women are associated with exposure to Diethylstilbestrol (DES) in the womb due to DES used as a medical treatment.

In another case, phthalates in pregnant women's urine was linked to subtle, but specific, genital changes in their male infants – a shorter, more female-like anogenital distance and associated incomplete descent of testes and a smaller scrotum and penis. The science behind this study has been questioned by phthalate industry consultants. As of June 2008, there are only five studies of anogenital distance in humans, and one researcher has stated "Whether AGD measures in humans relate to clinically important outcomes, however, remains to be determined, as does its utility as a measure of androgen action in epidemiologic studies."

U-shaped Dose-response Curve

Most toxicants, including endocrine disruptors, follow a U-shaped dose response curve. This means that very low and very high levels have more effects than mid-level exposure to a toxicant. Endocrine disrupting effects have been noted in animals exposed to environmentally relevant levels of some chemicals. For example, a common flame retardant, BDE-47, affects the reproductive system and thyroid gland of female rats in doses of the order of those to which humans are exposed. Low concentrations of endocrine disruptors can also have synergistic effects in amphibians, but it is not clear that this is an effect mediated through the endocrine system.

Critics have argued that data suggest that the amounts of chemicals in the environment are too low to cause an effect. A consensus statement by the Learning and Developmental Disabilities Initiative argued that "The very low-dose effects of endocrine disruptors cannot be predicted from high-dose studies, which contradicts the standard 'dose makes the poison' rule of toxicology. Nontraditional dose-response curves are referred to as nonmonotonic dose response curves."

The dosage objection could also be overcome if low concentrations of different endocrine disruptors are synergistic. This paper was published in Science in June 1996, and was one reason for the passage of the Food Quality Protection Act of 1996. The results could not be confirmed with the same and alternative methodologies, and the original paper was retracted, with Arnold found to have committed scientific misconduct by the United States Office of Research Integrity.

It has been claimed that Tamoxifen and some phthalates have fundamentally different (and harmful) effects on the body at low doses than at high doses.

Routes of Exposure

Food is a major mechanism by which people are exposed to pollutants. Diet is thought to account for up to 90% of a person's PCB and DDT body burden. In a study of 32 different common food products from three grocery stores in Dallas, fish and other animal products were found to be contaminated with PBDE. Since these compounds are fat soluble, it is likely they are accumulating from the environment in the fatty tissue of animals we eat. Some suspect fish consumption is a major source of many environmental contaminates. Indeed, both wild and farmed salmon from all over the world have been shown to contain a variety of man-made organic compounds.

With the increase in household products containing pollutants and the decrease in the quality of building ventilation, indoor air has become a significant source of pollutant exposure. Residents living in homes with wood floors treated in the 1960s with PCB-based wood finish have a much higher body burden than the general population. A study of indoor house dust and dryer lint of 16 homes found high levels of all 22 different PBDE congeners tested for in all samples. Recent studies suggest that contaminated house dust, not food, may be the major source of PBDE in our bodies. One study estimated that ingestion of house dust accounts for up to 82% of our PBDE body burden.

Research conducted by the Environmental Working Group found that 19 out of 20 children tested had levels of PBDE in their blood 3.5 times higher than the amount in their mothers' blood. It has been shown that contaminated housedust is a primary source of lead in young children's bodies. It may be that babies and toddlers ingest more contaminated housedust than the adults they live with, and therefore have much higher levels of pollutants in their systems.

Consumer goods are another potential source of exposure to endocrine disruptors. An analysis of the composition of 42 household cleaning and personal care products versus 43 "chemical free" products has been performed. The products contained 55 different chemical compounds: 50 were found in the 42 conventional samples representing 170 product types, while 41 were detected in 43 "chemical free" samples representing 39 product types. Parabens, a class of chemicals that has been associated with reproductive-tract issues, were detected in seven of the "chemical free" products, including three sunscreens that did not list parabens on the label. Vinyl products such as shower curtains were found to contain more than 10% by weight of the compound DEHP, which when present in dust has been associated with asthma and wheezing in children. The risk of exposure to EDCs increases as products, both conventional and "chemical free," are used in combination. "If a consumer used the alternative surface cleaner, tub and tile cleaner, laundry detergent, bar soap, shampoo and conditioner, facial cleanser and lotion, and toothpaste [he or she] would potentially be exposed to at least 19 compounds: 2 parabens, 3 phthalates, MEA, DEA, 5 alkylphenols, and 7 fragrances."

An analysis of the endocrine disrupting chemicals in Old Order Mennonite women in mid-pregnancy determined that they have much lower levels in their systems than the general population. Mennonites eat mostly fresh, unprocessed foods, farm without pesticides, and use few or no cosmetics or personal care products. One woman who had reported using hairspray and perfume had high levels of monoethyl phthalate, while the other women all had levels below detection. Three women who reported being in a car or truck within 48 hours of providing a urine sample had higher levels of diethylhexyl phthalate which is found in polyvinyl chloride, and is used in car interiors.

Additives added to plastics during manufacturing may leach into the environment after the plastic item is discarded; additives in microplastics in the ocean leach into ocean water and in plastics in landfills may escape and leach into the soil and then into groundwater.

Types

All people are exposed to chemicals with estrogenic effects in their everyday life, because endocrine disrupting chemicals are found in low doses in thousands of products. Chemicals commonly detected in people include DDT, polychlorinated biphenyls (PCB's), bisphenol A (BPA), polybrominated diphenyl ethers (PBDE's), and a variety of phthalates. In fact, almost all plastic products, including those advertised as "BPA free", have been found to leach endocrine-disrupting chemicals. In a 2011, study it was found that some "BPA-free" products released more endocrine active chemicals than the BPA-containing products.

There is some dispute in the scientific community surrounding the claim that these chemicals actually disrupt the endocrine system. Many believe that there is little evidence that the degree of exposure in humans is enough to warrant concern, while many others believe there is evidence that these chemicals pose some risk to human health.

Other forms of endocrine disruptors are phytoestrogens (plant hormones).

Some researchers are investigating the health risks to children of endocrine disrupting chemicals. Bisphenol A, until 2010 a common component in the plastic used to manufacture plastic baby bottles, has been banned in most countries. In 2010, despite strong industry opposition, Canada was the first to ban BPA's use in baby bottles. Australia and the European Union followed in 2011. Several states in the United States had banned its use by 2011, and in 2012 a nationwide ban was put in place.

Xenoestrogens

Xenoestrogens are a type of xenohormone that imitates estrogen. Synthetic xenoestrogens include widely used industrial compounds, such as PCBs, BPA and phthalates, which have estrogenic effects on a living organism.

Alkylphenols

Alkylphenols are xenoestrogens. The European Union has implemented sales and use restrictions on certain applications in which nonylphenols are used because of their alleged "toxicity, persistence, and the liability to bioaccumulate" but the United States EPA has taken a slower approach to make sure that action is based on "sound science".

Bisphenol A (BPA)

Bisphenol A is commonly found in plastic bottles, plastic food containers, dental materials, and the linings of metal food and infant formula cans. Another exposure comes from receipt paper commonly used at grocery stores and restaurants, because today the paper is commonly coated with a BPA containing clay for printing purposes.

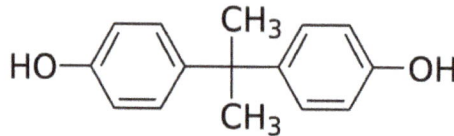

Bisphenol A chemical structure

BPA is a known endocrine disruptor, and numerous studies have found that laboratory animals exposed to low levels of it have elevated rates of diabetes, mammary and prostate cancers, decreased sperm count, reproductive problems, early puberty, obesity, and neurological problems. Early developmental stages appear to be the period of greatest sensitivity to its effects, and some studies have linked prenatal exposure to later physical and neurological difficulties. Regulatory bodies have determined safety levels for humans, but those safety levels are currently being questioned or are under review as a result of new scientific studies. A 2011 study that investigated the number of chemicals pregnant women are exposed to in the U.S. found BPA in 96% of women.

In 2010 the World Health Organization expert panel recommended no new regulations limiting or banning the use of Bisphenol-A, stating that "initiation of public health measures would be premature."

In August 2008, the U.S. FDA issued a draft reassessment, reconfirming their initial opinion that, based on scientific evidence, it is safe. However, in October 2008, FDA's advisory Science Board concluded that the Agency's assessment was "flawed" and hadn't proven the chemical to be safe for formula-fed infants. In January 2010, the FDA issued a report indicating that, due to findings of recent studies that used novel approaches in testing for subtle effects, both the National Toxicology Program at the National Institutes of Health as well as the FDA have some level of concern regarding the possible effects of BPA on the brain and behavior of fetuses, infants and younger children. In 2012 the FDA did ban the use of BPA in baby bottles, however the Environmental Working Group called the ban "purely cosmetic". In a statement they said,"If the agency truly wants to prevent people from being exposed to this toxic chemical associated with a variety of serious and chronic conditions it should ban its use in cans of infant formula, food and beverages." The Natural Resources Defense Council called the move inadequate saying, the FDA needs to ban BPA from all food packaging. In a statement a FDA spokesman said the agency's action was not based on safety concerns and that "the agency continues to support the safety of BPA for use in products that hold food."

Bisphenol S (BPS)

Bisphenol S is an analog of Bisphenol A. It is commonly found in thermal receipts, plastics, and household dust. Traces of BPS have also been found in personal care products. It is more presently being used because of the ban of BPA. BPS is used in place of BPA in "BPA free" items. However BPS has been shown to be as much of an endocrine disruptor as BPA.

Uses of long-chain Alkylphenols

The long-chain alkylphenols are used extensively as precursors to the detergents, as additives for fuels and lubricants, polymers, and as components in phenolic resins. These compounds are also used as building block chemicals that are also used in making fragrances, thermoplastic elastomers, antioxidants, oil field chemicals and fire retardant materials. Through the downstream use in making alkylphenolic resins, alkylphenols are also found in tires, adhesives, coatings, carbonless copypaper and high performance rubber products. They have been used in industry for over 40 years.

Certain alkylphenols are degradation products from nonionic detergents. Nonylphenol is considered to be a low-level endocrine disruptor owing to its tendency to mimic estrogen.

DDT

Dichloro-diphenyl-trichloroethane (DDT) was first used as a pesticide against Colorado potato beetles on crops beginning in 1936. An increase in the incidence of malaria, epidemic typhus, dysentery, and typhoid fever led to its use against the mosquitoes, lice, and houseflies that carried these diseases. Before World War II, pyrethrum, an extract of a flower from Japan, had been used to control these insects and the diseases they can spread. During World War II, Japan stopped exporting pyrethrum, forcing the search for an alternative. Fearing an epidemic outbreak of typhus, every British and American soldier was issued DDT, who used it to routinely dust beds, tents, and barracks all over the world.

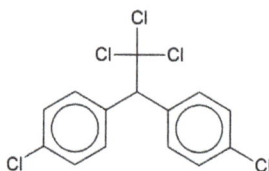
DDT Chemical structure

DDT was approved for general, non-military use after the war ended. It became used worldwide to increase monoculture crop yields that were threatened by pest infestation, and to reduce the spread of malaria which had a high mortality rate in many parts of the world. Its use for agricultural purposes has since been prohibited by national legislation of most countries, while its use as a control against malaria vectors is permitted, as specifically stated by the Stockholm Convention on Persistent Organic Pollutants

As early as 1946, the harmful effects of DDT on bird, beneficial insects, fish, and marine invertebrates were seen in the environment. The most infamous example of these effects were seen in the eggshells of large predatory birds, which did not develop to be thick enough to support the adult bird sitting on them. Further studies found DDT in high concentrations in carnivores all over the world, the result of biomagnification through the food chain. Twenty years after its widespread use, DDT was found trapped in ice samples taken from Antarctic snow, suggesting wind and water are another means of environmental transport. Recent studies show the historical record of DDT deposition on remote glaciers in the Himalayas.

More than sixty years ago when biologists began to study the effects of DDT on laboratory animals, it was discovered that DDT interfered with reproductive development. Recent studies suggest DDT may inhibit the proper development of female reproductive organs that adversely affects reproduction into maturity. Additional studies suggest that a marked decrease in fertility in adult males may be due to DDT exposure. Most recently, it has been suggested that exposure to DDT in utero can increase a child's risk of childhood obesity. DDT is still used as anti-malarial insecticide in Africa and parts of Southeast Asia in limited quantities.

Polychlorinated Biphenyls

Polychlorinated biphenyls (PCBs) are a class of chlorinated compounds used as industrial coolants and lubricants. PCBs are created by heating benzene, a byproduct of gasoline refining, with chlorine. They were first manufactured commercially by the Swann Chemical Company in 1927. In 1933, the health effects of direct PCB exposure was seen in those who worked with the chemicals at the manufacturing facility in Alabama. In 1935, Monsanto acquired the company, taking over US production and licensing PCB manufacturing technology internationally.

General Electric was one of the largest US companies to incorporate PCBs into manufactured equipment. Between 1952 and 1977, the New York GE plant had dumped more than 500,000 pounds of PCB waste into the Hudson River. PCBs were first discovered in the environment far from its industrial use by scientists in Sweden studying DDT.

The effects of acute exposure to PCBs were well known within the companies who used Monsanto's PCB formulation who saw the effects on their workers who came into contact with it regularly. Direct skin contact results in a severe acne-like condition called chloracne. Exposure increases the risk of skin cancer, liver cancer, and brain cancer. Monsanto tried for years to downplay the health problems related to PCB exposure in order to continue sales.

The detrimental health effects of PCB exposure to humans became undeniable when two separate incidents of contaminated cooking oil poisoned thousands of residents in Japan (Yushō disease, 1968) and Taiwan (Yu-cheng disease, 1979), leading to a worldwide ban on PCB use in 1977. Recent studies show the endocrine interference of certain PCB congeners is toxic to the liver and thyroid, increases childhood obesity in children exposed prenatally, and may increase the risk of developing diabetes.

PCBs in the environment may also be related to reproductive and infertility problems in wildlife. In Alaska it is thought that they may contribute to reproductive defects, infertility and antler malformation in some deer populations. Declines in the populations of otters and sea lions may also be partially due to their exposure to PCBs, the insecticide DDT, other persistent organic pollutants. Bans and restrictions on the use of EDCs have been associated with a reduction in health problems and the recovery of some wildlife populations.

Polybrominated Diphenyl Ethers

Polybrominated diphenyl ethers (PBDEs) are a class of compounds found in flame retardants used in plastic cases of televisions and computers, electronics, carpets, lighting, bedding, clothing, car components, foam cushions and other textiles. Potential health concern: PBDE's are structurally very similar to Polychlorinated biphenyls (PCBs), and have similar neurotoxic effects. Research has correlated halogenated hydrocarbons, such as PCBs, with neurotoxicity. PBDEs are similar in chemical structure to PCBs, and it has been suggested that PBDEs act by the same mechanism as PCBs.

In the 1930s and 1940s, the plastics industry developed technologies to create a variety of plastics with broad applications. Once World War II began, the US military used these new plastic materials to improve weapons, protect equipment, and to replace heavy components in aircraft and vehicles. After WWII, manufacturers saw the potential plastics could have in many industries, and plastics were incorporated into new consumer product designs. Plastics began to replace wood and metal in existing products as well, and today plastics are the most widely used manufacturing materials.

By the 1960s, all homes were wired with electricity and had numerous electrical appliances. Cotton had been the dominant textile used to produce home furnishings, but now home furnishings were composed of mostly synthetic materials. More than 500 billion cigarettes were consumed each year in the 1960s, as compared to less than 3 billion per year in the beginning of the twentieth century. When combined with high density living, the potential for home fires was higher in the 1960s than it had ever been in the US. By the late 1970s, approximately 6000 people in the US died each year in home fires.

In 1972, in response to this situation, the National Commission on Fire Prevention and Control was created to study the fire problem in the US. In 1973 they published their findings in America Burning, a 192-page report that made recommendations to increase fire prevention. Most of the recommendations dealt with fire prevention education and improved building engineering, such as the installation of fire sprinklers and smoke detectors. The Commission expected that with the recommendations, a 5% reduction in fire losses could be expected each year, halving the annual losses within 14 years.

Historically, treatments with alum and borax were used to reduce the flammability of fabric and wood, as far back as Roman times. Since it is a non-absorbent material once created, flame retardant chemicals are added to plastic during the polymerization reaction when it is formed. Organic compounds based on halogens like bromine and chlorine are used as the flame retardant additive in plastics, and in fabric based textiles as well. The widespread use of brominated flame retardants may be due to the push from Great Lakes Chemical Corporation (GLCC) to profit from its huge investment in bromine. In 1992, the world market consumed approximately 150,000 tonnes of bromine-based flame retardants, and GLCC produced 30% of the world supply.

PBDEs have the potential to disrupt thyroid hormone balance and contribute to a variety of neurological and developmental deficits, including low intelligence and learning disabilities. Many of the most common PBDE's were banned in the European Union in 2006. Studies with rodents have suggested that even brief exposure to PBDEs can cause developmental and behavior problems in juvenile rodents and exposure interferes with proper thyroid hormone regulation.

Phthalates

Phthalates are found in some soft toys, flooring, medical equipment, cosmetics and air fresheners. They are of potential health concern because they are known to disrupt the endocrine system of animals, and some research has implicated them in the rise of birth defects of the male reproductive system.

Although an expert panel has concluded that there is "insufficient evidence" that they can harm the reproductive system of infants, California and Europe have banned them from toys. One phthalate, Bis(2-ethylhexyl) phthalate (DEHP), used in medical tubing, catheters and blood bags, may harm sexual development in male infants. In 2002, the Food and Drug Administration released a public report which cautioned against exposing male babies to DEHP. Although there are no direct human studies the FDA report states: "Exposure to DEHP has produced a range of adverse effects in laboratory animals, but of greatest concern are effects on the development of the male reproductive system and production of normal sperm in young animals. In view of the available animal data, precautions should be taken to limit the exposure of the developing male to DEHP". Similarly, phthalates may play a causal role in disrupting masculine neurological development when exposed prenatally.

Perfluorooctanoic Acid

PFOA exerts hormonal effects including alteration of thyroid hormone levels. Blood serum levels of PFOA were associated with an increased time to pregnancy — or "infertility" — in a 2009 study. PFOA exposure is associated with decreased semen quality. PFOA appeared to act as an endocrine disruptor by a potential mechanism on breast maturation in young girls. A C8 Science Panel status report noted an association between exposure in girls and a later onset of puberty.

Other Suspected Endocrine Disruptors

Some other examples of putative EDCs are polychlorinated dibenzo-dioxins (PCDDs) and -furans (PCDFs), polycyclic aromatic hydrocarbons (PAHs), phenol derivatives and a number of pesticides (most prominent being organochlorine insecticides like endosulfan, Kepone(chlordecone) and DDT and its derivatives, the herbicide atrazine, and the fungicide vinclozolin), the contraceptive 17-alpha ethinylestradiol, as well as naturally occurring phytoestrogens such as genistein and mycoestrogens such as zearalenone.

The molting in crustaceans is an endocrine-controlled process. In the marine penaeid shrimp *Litopenaeus vannamei*, exposure to endosulfan resulted increased susceptibility to acute toxicity and increased mortalities in the postmolt stage of the shrimp.

Many sunscreens contain oxybenzone, a chemical blocker that provides broad-spectrum UV coverage, yet is subject to a lot of controversy due its potential estrogenic effect in humans.

Tributyltin (TBT) are organotin compounds that for 40 years TBT was used as a biocide in anti-fouling paint, commonly known as bottom paint. TBT has been shown to impact invertebrate and vertebrate development, disrupting the endorcrine system, resulting in masculinization, lower survival rates,as well as many health problems in mammals.

Temporal Trends of Body Burden

Since being banned, the average human body burdens of DDT and PCB have been declining. Since their ban in 1972, the PCB body burden is 1/100 of what it was in the early 1980s (Weschler 2009). Monitoring programs of European breast milk samples have shown that PBDE levels are increasing. An analysis of PBDE content in breast milk samples from Europe, Canada, and the US shows that levels are 40 times higher for North American women than for Swedish women, and that levels in North America are doubling every two to six years.

Legal Approach

United States

The multitude of possible endocrine disruptors are technically regulated in the United States by many laws, including: the Toxic Substances Control Act, the Federal Insecticide, Fungicide, and Rodenticide Act, the Food, Drug and Cosmetic Act, the Clean Water Act, the Safe Drinking Water Act, and the Clean Air Act.

The Congress of the United States has improved the evaluation and regulation process of drugs and other chemicals. The Food Quality Protection Act of 1996 and the Safe Drinking Water Act of 1996 simultaneously provided the first legislative direction requiring the EPA to address endocrine disruption through establishment of a program for screening and testing of chemical substances.

In 1998 the EPA announced the Endocrine Disruptor Screening Program by establishment of a framework for priority setting, screening and testing more than 85,000 chemicals in commerce. The basic concept behind the program is that prioritization will be based on existing information about chemical uses, production volume, structure-activity and toxicity. Screening is done by use of *in vitro* test systems (by examining, for instance, if an agent interacts with the estrogen receptor or the androgen receptor) and via the use of in animal models, such as development of tadpoles and uterine growth in prepubertal rodents. Full scale testing will examine effects not only in mammals (rats) but also in a number of other species (frogs, fish, birds and invertebrates). Since the theory involves the effects of these substances on a functioning system, animal testing is essential for scientific validity, but has been opposed by animal rights groups. Similarly, proof that these effects occur in humans would require human testing, and such testing also has opposition.

After failing to meet several deadlines to begin testing, the EPA finally announced that they were ready to begin the process of testing dozens of chemical entities that are suspected endocrine disruptors early in 2007, eleven years after the program was announced. When the final structure of the tests was announced there was objection to their design. Critics have charged that the entire process has been compromised by chemical company interference. In 2005, the EPA appointed a panel of experts to conduct an open peer-review of the program and its orientation. Their results found that "the long-term goals and science questions in the EDC program are appropriate", however this study was conducted over a year before the EPA announced the final structure of the screening program.

Europe

In 2013, a number of pesticides containing endocrine disrupting chemicals were in draft EU criteria to be banned. On the 2nd May, US TTIP negotiators insisted the EU drop the criteria. They stated that a risk-based approach should be taken on regulation. Later the same day Catherine Day wrote to Karl Falkenberg asking for the criteria to be removed.

The European Commission had been to set criteria by December 2013 identifying endocrine disrupting chemicals (EDCs) in thousands of products — including disinfectants, pesticides and toiletries — that have been linked to cancers, birth defects and development disorders in children. However, the body delayed the process, prompting Sweden to state that it would sue the commission in May 2014 — blaming chemical industry lobbying for the disruption.

"This delay is due to the European chemical lobby, which put pressure again on different commissioners. Hormone disrupters are becoming a huge problem. In some places in Sweden we see double-sexed fish. We have scientific reports on how this affects fertility of young boys and girls, and other serious effects," Swedish Environment Minister Lena Ek told the AFP, noting that Denmark had also demanded action.

In November 2014, the Copenhagen-based Nordic Council of Ministers released its own independent report that estimated the impact of environmental EDCs on male reproductive health, and the resulting cost to public health systems. It concluded that EDCs likely cost health systems across the EU anywhere from 59 million to 1.18 billion Euros a year, noting that even this represented only "a fraction of the endocrine related diseases".

Environmental and Human Body Cleanup

There is evidence that once a pollutant is no longer in use, or once its use is heavily restricted, the human body burden of that pollutant declines. Through the efforts of several large-scale monitoring programs, the most prevalent pollutants in the human population are fairly well known. The first step in reducing the body burden of these pollutants is eliminating or phasing out their production.

The second step toward lowering human body burden is awareness of and potentially labeling foods that are likely to contain high amounts of pollutants. This strategy has worked in the past - pregnant and nursing women are cautioned against eating sea-food that is known to accumulate high levels of mercury. Ideally, a certification process should be in place to routinely test animal products for POP concentrations. This would help the consumer identify which foods have the highest levels of pollutants.

The most challenging aspect of this problem is discovering how to eliminate these compounds from the environment and where to focus remediation efforts. Even pollutants no longer in production persist in the environment, and bio-accumulate in the food chain. An understanding of how these chemicals, once in the environment, move through ecosystems, is essential to designing ways to isolate and remove them. Working backwards through the food chain may help to identify areas to prioritize for remediation efforts. This may be extremely challenging for contaminated fish and marine mammals that have a large habitat and who consume fish from many different areas throughout their lives.

Many persistent organic compounds, PCB, DDT and PBDE included, accumulate in river and marine sediments. Several processes are currently being used by the EPA to clean up heavily polluted areas, as outlined in their Green Remediation program.

One of the most interesting ways is the utilization of naturally occurring microbes that degrade PCB congeners to remediate contaminated areas.

There are many success stories of cleanup efforts of large heavily contaminated Super-fund sites. A 10-acre (40,000 m²) landfill in Austin, Texas contaminated with illegally dumped VOCs was restored in a year to a wetland and educational park.

A US uranium enrichment site that was contaminated with uranium and PCBs was cleaned up with high tech equipment used to find the pollutants within the soil. The soil and water at a polluted wetlands site were cleaned of VOCs, PCBs and lead, native plants were installed as biological filters, and a community program was implemented to ensure ongoing monitoring of pollutant concentrations in the area. These case studies are encouraging due to the short amount of time needed to remediate the site and the high level of success achieved.

Studies suggest that bisphenol A, certain PCBs, and phthalate compounds are preferentially eliminated from the human body through sweat.

Economic Effects

Human exposure may cause some health effects, such as lower IQ and adult obesity. These effects may lead to lost productivity, disability, or premature death in some people. One source estimated that, within the European Union, this economic effect might have about twice the economic impact as the effects caused by mercury and lead contamination.

The socio-economic burden of EDC-associated health effects for the EU was estimated based on currently available literature and considering the uncertainties with respect to causality with EDCs and corresponding health-related costs to be in the range of 46 and 288 Euros per year.

Neurotoxin

Neurotoxins are toxins that are poisonous or destructive to nerve tissue (causing neurotoxicity). Neurotoxins are an extensive class of exogenous chemical neurological insults that can adversely affect function in both developing and mature nervous tissue. The term can also be used to classify endogenous compounds, which, when abnormally contact , can prove neurologically toxic. Though neurotoxins are often neurologically destructive, their ability to specifically target neural components is important in the study of nervous systems. Common examples of neurotoxins include lead, ethanol (drinking alcohol), manganese glutamate, nitric oxide (NO), botulinum toxin (e.g. Botox), tetanus toxin, and tetrodotoxin. Some substances such as nitric oxide and glutamate are in fact essential for proper function of the body and only exert neurotoxic effects at excessive concentrations.

Neurotoxins can be found in a number of organisms, including some strains of cyanobacteria, that can be found in algal blooms or washed up on shore in a green scum.

Neurotoxins inhibit neuron control over ion concentrations across the cell membrane, or communication between neurons across a synapse. Local pathology of neurotoxin exposure often includes neuron excitotoxicity or apoptosis but can also include glial cell damage. Macroscopic manifestations of neurotoxin exposure can include widespread central nervous system damage such as intellectual disability, persistent memory impairments, epilepsy, and dementia. Additionally, neurotoxin-mediated peripheral nervous system damage such as neuropathy or myopathy is common. Support has been shown for a number of treatments aimed at attenuating neurotoxin-mediated injury, such as antioxidant and antitoxin administration.

Background

Exposure to neurotoxins in society is not new, as civilizations have been exposed to neurologically destructive compounds for thousands of years. One notable example is the possible significant lead exposure during the Roman Empire resulting from the development of extensive plumbing networks and the habit of boiling vinegared wine in lead pans to sweeten it, the process generating lead acetate, known as "sugar of lead". In part, neurotoxins have been part of human history because of the fragile and susceptible nature of the nervous system, making it highly prone to disruption.

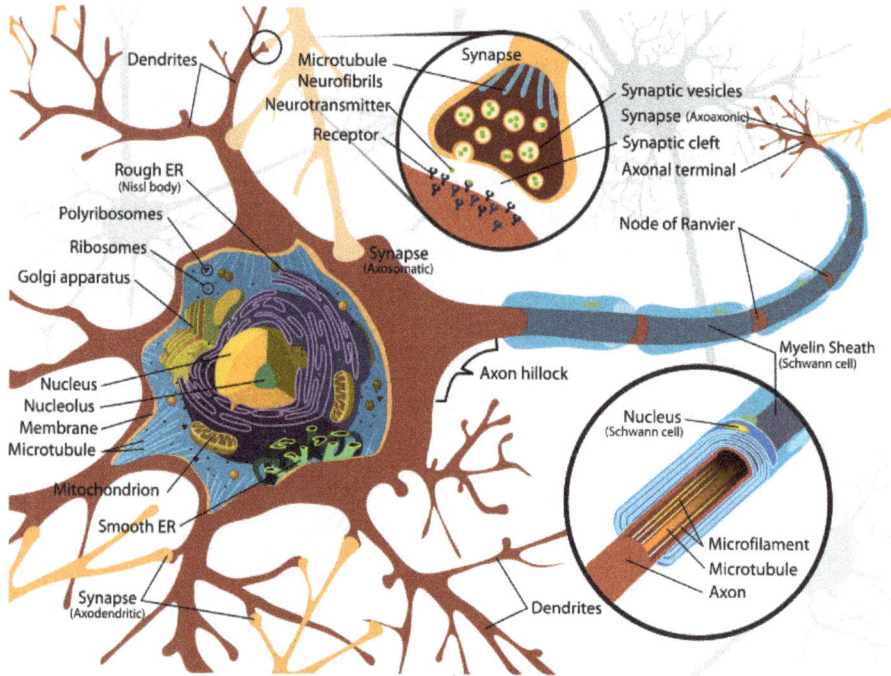

Illustration of typical multipolar neuron

The nervous tissue found in the brain, spinal cord, and periphery comprises an extraordinarily complex biological system that largely defines many of the unique traits of individuals. As with any highly complex system, however, even small perturbations to its environment can lead to significant functional disruptions. Properties leading to the susceptibility of nervous tissue include a high surface area of neurons, a high lipid content which retains lipophilic toxins, high blood flow to the brain inducing increased effective toxin exposure, and the persistence of neurons through an individual's lifetime, leading to compounding of damages. As a result, the nervous system has a number of mechanisms designed to protect it from internal and external assaults, including the blood brain barrier.

The blood-brain barrier (BBB) is one critical example of protection which prevents toxins and other adverse compounds from reaching the brain. As the brain requires nutrient entry and waste removal, it is perfused by blood flow. Blood can carry a number of ingested toxins, however, which would induce significant neuron death if they reach nervous tissue. Thus, protective cells termed astrocytes surround the capillaries in the brain and absorb nutrients from the blood and subsequently transport them to the neurons, effectively isolating the brain from a number of potential chemical insults.

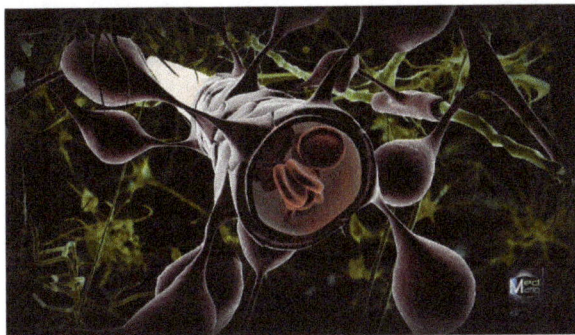

Astrocytes surrounding capillaries in the brain to form the blood brain barrier

This barrier creates a tight hydrophobic layer around the capillaries in the brain, inhibiting the transport of large or hydrophilic compounds. In addition to the BBB, the choroid plexus provides a layer of protection against toxin absorption in the brain. The choroid plexuses are vascularized layers of tissue found in the third, fourth, and lateral ventricles of the brain, which through the function of their ependymal cells, are responsible for the synthesis of cerebrospinal fluid (CSF). Importantly, through selective passage of ions and nutrients and trapping heavy metals such as lead, the choroid plexuses maintain a strictly regulated environment which contains the brain and spinal cord.

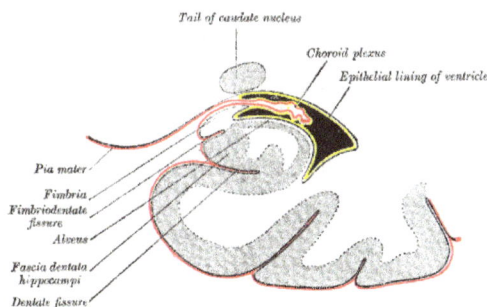

Choroid plexus

By being hydrophobic and small, or inhibiting astrocyte function, some compounds including certain neurotoxins are able to penetrate into the brain and induce significant damage. In modern times, scientists and physicians have been presented with the challenge of identifying and treating neurotoxins, which has resulted in a growing interest in both neurotoxicology research and clinical studies. Though clinical neurotoxicology is largely a burgeoning field, extensive inroads have been made in the identification of many environmental neurotoxins leading to the classification of 750 to 1000 known potentially neurotoxic compounds. Due to the critical importance of finding neurotoxins in common environments, specific protocols have been developed by the United States Environmental Protection Agency (EPA) for testing and determining neurotoxic effects of compounds (USEPA 1998). Additionally, in vitro systems have increased in use as they provide significant improvements over the more common in vivo systems of the past. Examples of improvements include tractable, uniform environments, and the elimination of contaminating effects of systemic metabolism. In vitro systems, however, have presented problems as it has been difficult to properly replicate the complexities of the nervous system, such as the interactions between supporting astrocytes and neurons in creating the BBB. To even further complicate the process of determining neurotoxins when testing in-vitro, neurotoxicity and cytotoxicity may be difficult to distinguish as exposing neurons directly to compounds may not be possible in-vivo, as it is in-vitro. Additionally, the response of cells to chemicals may not accurately convey a distinction between neurotoxins and cytotoxins, as symptoms like oxidative stress or skeletal modifications may occur in response to either.

In an effort to address this complication, neurite outgrowths (either axonal or dendritic) in response to applied compounds have recently been proposed as a more accurate distinction between true neurotoxins and cytotoxins in an in-vitro testing environment. Due to the significant inaccuracies associated with this process, however, it has been slow in gaining widespread support. Additionally, biochemical mechanisms have become more widely used in neurotoxin testing, such that compounds can be screened for sufficiency to induce cell mechanism interference, like the inhibition of acetylcholinesterase capacity of organophosphates (includes DDT and sarin gas). Though methods of determining neurotoxicity still require significant development, the identification of deleterious compounds and toxin exposure symptoms has undergone significant improvement.

Applications in Neuroscience

Though diverse in chemical properties and functions, neurotoxins share the common property that they act by some mechanism leading to either the disruption or destruction of necessary components within the nervous system. Neurotoxins, however, by their very design can be very useful in the field of neuroscience. As the nervous system in most organisms is both highly complex and necessary for survival, it has naturally become a target for attack by both predators and prey. As venomous organisms often use their neurotoxins to subdue a predator or prey very rapidly, toxins have evolved to become highly specific to their target channels such that the toxin does not readily bind other targets. As such, neurotoxins provide an effective means by which certain elements of the nervous system may be accurately and efficiently targeted. An early example of neurotoxin based targeting used radiolabeled tetrodotoxin to assay sodium channels and obtain precise measurements about their concentration along nerve membranes. Likewise through isolation of certain channel activities, neurotoxins have provided the ability to improve the original Hodgkin-Huxley model of the neuron in which it was theorized that single generic sodium and potassium channels could account for most nervous tissue function. From this basic understanding, the use of common compounds such as tetrodotoxin, tetraethylammonium, and bungarotoxins have led to a much deeper understanding of the distinct ways in which individual neurons may behave.

Mechanisms of Activity

As neurotoxins are compounds which adversely affect the nervous system, a number of mechanisms through which they function are through the inhibition of neuron cellular processes. These inhibited processes can range from membrane depolarization mechanisms to inter-neuron communication. By inhibiting the ability for neurons to perform their expected intracellular functions, or pass a signal to a neighboring cell, neurotoxins can induce systemic nervous system arrest as in the case of botulinum toxin, or even nervous tissue death. The time required for the onset of symptoms upon neurotoxin exposure can vary between different toxins, being on the order of hours for botulinum toxin and years for lead.

Neurotoxin classification	Neurotoxins
Na channel inhibitors	Tetrodotoxin
K channel inhibitors	Tetraethylammonium
Cl channel inhibitors	Chlorotoxin,
Ca channel inhibitors	Conotoxin
Inhibitors of synaptic vesicle release	Botulinum toxin, tetanus toxin
Receptor inhibitors	Bungarotoxin Curare
Receptor agonists	25I-NBOMe JWH-018
Blood brain barrier inhibitors	Aluminium, mercury
Cytoskeleton interference	Arsenic, ammonia
Ca-mediated cytotoxicity	Lead
Multiple effects	Ethanol
Endogenous neurotoxin sources	Nitric oxide, Glutamate, Dopamine

Inhibitors

Tetrodotoxin

Tetrodotoxin (TTX) is a poison produced by organisms belonging to the Tetradontidae order, which includes the puffer fish, ocean sunfish, and porcupine fish. Within the puffer fish, which is a common delicacy especially in Japan, TTX is found in the liver, gonads, ovaries, intestines, and skin. TTX can be fatal if consumed, and has become a common form of poisoning in many countries. Common symptoms of TTX consumption include paraesthesia (often restricted to the mouth and limbs), muscle weakness, nausea, and vomiting and often manifest within 30 minutes of ingestion. The primary mechanism by which TTX is toxic is through the inhibition of sodium channel function, which reduces the functional capacity of neuron communication. This inhibition largely affects a susceptible subset of sodium channels known as TTX-sensitive (TTX-s), which also happens to be largely responsible for the sodium current that drives the depolarization phase of neuron action potentials.

The puffer fish is a well known tetrodotoxin producer.

TTX-resistant (TTX-r) is another form of sodium channel which has limited sensitivity to TTX, and is largely found in small diameter axons such as those found in nociception neurons. When significant levels of TTX is ingested, it will bind sodium channels on neurons and reduce their membrane permeability to sodium. This results in an increased effective threshold of required excitatory signals in order to induce an action potential in a postsynaptic neuron. The effect of this increased signaling threshold is a reduced excitability of postsynaptic neurons, and subsequent loss of motor and sensory function which can result in paralysis and death. Though assisted ventilation may increase the chance of survival after TTX exposure, there is currently no antitoxin. The use of the acetylcholinesterase inhibitor Neostigmine or the muscarinic acetylcholine antagonist Atropine (which will inhibit parasympathetic activity), however, can increase sympathetic nerve activity enough to improve the chance of survival after TTX exposure.

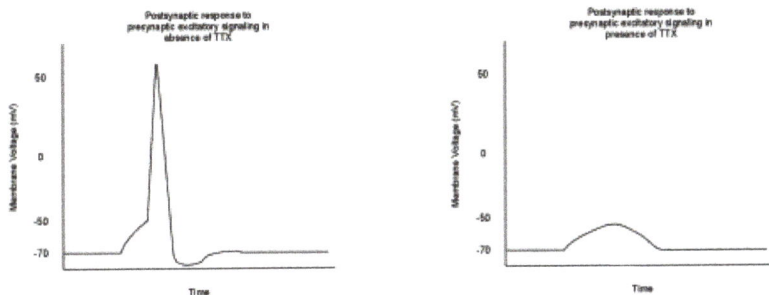

Inhibited signaling response resulting from neuron exposure to tetrodotoxin.

Potassium Channel

Tetraethylammonium

Tetraethylammonium (TEA) is a compound that, like a number of neurotoxins, was first identified through its damaging effects to the nervous system and shown to have the capacity of inhibiting the function of motor nerves and thus the contraction of the musculature in a manner similar to that of curare. Additionally, through chronic TEA administration, muscular atrophy would be induced. It was later determined that TEA functions in-vivo primarily through its ability to inhibit both the potassium channels responsible for the delayed rectifier seen in an action potential and some population of calcium-dependent potassium channels. It is this capability to inhibit potassium flux in neurons that has made TEA one of the most important tools in neuroscience. It has been hypothesized that the ability for TEA to inhibit potassium channels is derived from its similar space-filling structure to potassium ions. What makes TEA very useful for neuroscientists is its specific ability to eliminate potassium channel activity, thereby allowing the study of neuron response contributions of other ion channels such as voltage gated sodium channels. In addition to its many uses in neuroscience research, TEA has been shown to perform as an effective treatment of Parkinson's disease through its ability to limit the progression of the disease.

Chloride Channel

Chlorotoxin

Chlorotoxin (Cltx) is the active compound found in scorpion venom, and is primarily toxic because of its ability to inhibit the conductance of chloride channels. Ingestion of lethal volumes of Cltx results in paralysis through this ion channel disruption. Similar to botulinum toxin, Cltx has been shown to possess significant therapeutic value. Evidence has shown that Cltx can inhibit the ability for gliomas to infiltrate healthy nervous tissue in the brain, significantly reducing the potential invasive harm caused by tumors.

Calcium Channel

Conotoxin

Conotoxins represent a category of poisons produced by the marine cone snail, and are capable of inhibiting the activity of a number of ion channels such as calcium, sodium, or potassium channels. In many cases, the toxins released by the different types of cone snails include a range of different types of conotoxins, which may be specific for different ion channels, thus creating a venom capable of widespread nerve function interruption. One of the unique forms of conotoxins, ω-conotoxin (ω-CgTx) is highly specific for Ca channels and has shown usefulness in isolating them from a system. As calcium flux is necessary for proper excitability of a cell, any significant inhibition could prevent a large amount of functionality. Significantly, ω-CgTx is capable of long term binding to and inhibition of voltage-dependent calcium channels located in the membranes of neurons but not those of muscle cells.

Synaptic Vesicle Release

Botulinum Toxin

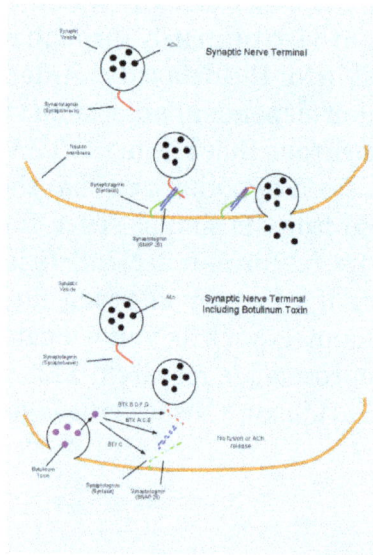

Mechanism of Botulinum Toxin neurotoxicity

Botulinum Toxin (BTX) is a group of neurotoxins consisting of eight distinct compounds, referred to as BTX-A,B,C,D,E,F,G,H, which are produced by the bacterium *Clostridium botulinum* and lead to muscular paralysis. A notably unique feature of BTX is its relatively common therapeutic use in treating dystonia and spasticity disorders, as well as in inducing muscular atrophy despite being the most poisonous substance known. BTX functions peripherally to inhibit acetylcholine (ACh) release at the neuromuscular junction through degradation of the SNARE proteins required for ACh vesicle-membrane fusion. As the toxin is highly biologically active, an estimated dose of 1μg/kg body weight is sufficient to induce an insufficient tidal volume and resultant death by asphyxiation. Due to its high toxicity, BTX antitoxins have been an active area of research. It has been shown that capsaicin (active compound responsible for heat in chili peppers) can bind the TRPV1 receptor expressed on cholinergic neurons and inhibit the toxic effects of BTX.

Tetanus Toxin

Tetanus neurotoxin (TeNT) is a compound that functionally reduces inhibitory transmissions in the nervous system resulting in muscular tetany. TeNT is similar to BTX, and is in fact highly similar in structure and origin; both belonging to the same category of clostridial neurotoxins. Like BTX, TeNT inhibits inter-neuron communication by means of vesicular neurotransmitter (NT) release. One notable difference between the two compounds is that while BTX inhibits muscular contractions, TeNT induces them. Though both toxins inhibit vesicle release at neuron synapses, the reason for this different manifestation is that BTX functions mainly in the peripheral nervous system (PNS) while TeNT is largely active in the central nervous system (CNS). This is a result of TeNT migration through motor neurons to the inhibitory neurons of the spinal cord after entering through endocytosis. This results in a loss of function in inhibitory neurons within the CNS resulting in systemic muscular contractions. Similar to the prognosis of a lethal dose of BTX, TeNT leads to paralysis and subsequent suffocation.

Blood Brain Barrier

Aluminium

Neurotoxic behavior of aluminum is known to occur upon entry into the circulatory system, where it can migrate to the brain and inhibit some of the crucial functions of the blood brain barrier (BBB). A loss of function in the BBB can produce significant damage to the neurons in the CNS, as the barrier protecting the brain from other toxins found in the blood will no longer be capable of such action. Though the metal is known to be neurotoxic, effects are usually restricted to patients incapable of removing excess ions from the blood, such as those experiencing renal failure. Patients experiencing aluminum toxicity can exhibit symptoms such as impaired learning and reduced motor coordination. Additionally, systemic aluminum levels are known to increase with age, and have been shown to correlate with Alzheimer's Disease, implicating it as a neurotoxic causative compound of the disease. Despite its known toxicity, aluminum is still extensively utilized in the packaging and preparing of food, while other toxic metals such as lead have been almost entirely phased-out of use in these industries.

Mercury

Mercury is capable of inducing CNS damage by migrating into the brain by crossing the BBB. Mercury exists in a number of different compounds, though methylmercury

(MeHg⁺), dimethylmercury and diethylmercury are the only significantly neurotoxic forms. Diethylmercury and dimethylmercury are considered some of the most potent neurotoxins ever discovered. MeHg⁺ is usually acquired through consumption of seafood, as it tends to concentrate in organisms high on the food chain. It is known that the mercuric ion inhibits amino acid (AA) and glutamate (Glu) transport, potentially leading to excitotoxic effects.

Receptor Agonists and Antagonists

Anatoxin-a

Investigations into anatoxin-*a*, also known as "Very Fast Death Factor", began in 1961 following the deaths of cows that drank from a lake containing an algal bloom in Saskatchewan, Canada. It is a cyanotoxin produced by at least four different genera of cyanobacteria, and has been reported in North America, Europe, Africa, Asia, and New Zealand.

External video
Very Fast Death Factor University of Nottingham

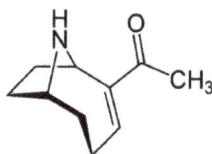

Anatoxin-*a*

Toxic effects from anatoxin-*a* progress very rapidly because it acts directly on the nerve cells (neurons). The progressive symptoms of anatoxin-*a* exposure are loss of coordination, twitching, convulsions and rapid death by respiratory paralysis. The nerve tissues which communicate with muscles contain a receptor called the nicotinic acetylcholine receptor. Stimulation of these receptors causes a muscular contraction. The anatoxin-*a* molecule is shaped so it fits this receptor, and in this way it mimics the natural neurotransmitter normally used by the receptor, acetylcholine. Once it has triggered a contraction, anatoxin-*a* does not allow the neurons to return to their resting state, because it is not degraded by cholinesterase which normally performs this function. As a result, the muscle cells contract permanently, the communication between the brain and the muscles is disrupted and breathing stops.

When it was first discovered, the toxin was called the Very Fast Death Factor (VFDF) because when it was injected into the body cavity of mice it induced tremors, paralysis and death within a few minutes. In 1977, the structure of VFDF was determined as a secondary, bicyclic amine alkaloid, and it was renamed anatoxin-*a*. Structurally, it is similar to cocaine. There is continued interest in anatoxin-*a* because of the dangers it presents to recreational and drinking waters, and because it is a particularly useful molecule for investigating acetylcholine receptors in the nervous system. The deadliness of the toxin means that it has a high military potential as a toxin weapon.

Bungarotoxin

Bungarotoxin is a compound with known interaction with nicotinic acetylcholine receptors (nAChRs), which constitute a family of ion channels whose activity is triggered by neurotransmitter binding. Bungarotoxin is produced in a number of different forms, though one of the commonly used forms is the long chain alpha form, α-bungarotoxin, which is isolated from the banded krait snake. Though extremely toxic if ingested, α-bungarotoxin has shown extensive usefulness in neuroscience as it is particularly adept at isolating nAChRs due to its high affinity to the receptors. As there are multiple forms of bungarotoxin, there are different forms of nAChRs to which they will bind, and α-bungarotoxin is particularly specific for α7-nAChR. This α7-nAChR functions to allow calcium ion influx into cells, and thus when blocked by ingested bungarotoxin will produce damaging effects, as ACh signaling will be inhibited. Likewise, the use of α-bungarotoxin can be very useful in neuroscience if it is desirable to block calcium flux in order to isolate effects of other channels. Additionally, different forms of bungarotoxin may be useful for studying inhibited nAChRs and their resultant calcium ion flow in different systems of the body. For example, α-bungarotoxin is specific for nAChRs found in the musculature and κ-bungarotoxin is specific for nAChRs found in neurons.

Caramboxin

Caramboxin (CBX) is a toxin found in star fruit (*Averrhoa carambola*). Individuals with some types of kidney disease are susceptible to adverse neurological effects including intoxication, seizures and even death after eating star fruit or drinking juice made of this fruit. Caramboxin is a new nonpeptide amino acid toxin that stimulate the glutamate receptors in neurons. Caramboxin is an agonist of both NMDA and AMPA glutamatergic ionotropic receptors with potent excitatory, convulsant, and neurodegenerative properties.

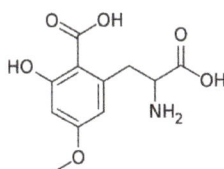

Caramboxin

Curare

The term "curare" is ambiguous because it has been used to describe a number of poisons which at the time of naming were understood differently from present day understandings. In the past the characterization has meant poisons used by South American tribes on arrows or darts, though it has matured to specify a specific categorization of poisons which act on the neuromuscular junction to inhibit signaling and thus induce muscle relaxation. The neurotoxin category contains a number of distinct poisons, though all were originally purified from plants originating in South America. The effect with which injected curare poison is usually associated is muscle paralysis and resultant death. Curare notably functions to inhibit nicotinic acetylcholine receptors at the neuromuscular junction. Normally, these receptor channels allow sodium ions into muscle cells to initiate an action potential that leads to muscle contraction. By blocking the receptors, the neurotoxin is capable of significantly reducing neuromuscular junction signaling, an effect which has resulted in its use by anesthesiologists to produce muscular relaxation.

Cytoskeleton Interference

Arsenic

Arsenic is a neurotoxin commonly found concentrated in areas exposed to agricultural runoff, mining, and smelting sites (Martinez-Finley 2011). One of the effects of arsenic ingestion during the development of the nervous system is the inhibition of neurite growth which can occur both in PNS and the CNS. This neurite growth inhibition can often lead to defects in neural migration, and significant morphological changes of neurons during development,) often leading to neural tube defects in neonates. As a metabolite of arsenic, arsenite is formed after ingestion of arsenic and has shown significant toxicity to neurons within about 24 hours of exposure. The mechanism of this cytotoxicity functions through arsenite-induced increases in intracellular calcium ion levels within neurons, which may subsequently reduce mitochondrial transmembrane potential which activates caspases, triggering cell death. Another known function of arsenite is its destructive nature towards the cytoskeleton through inhibition of neurofilament transport. This is particularly destructive as neurofilaments are used in basic cell structure and support. Lithium administion has shown promise, however, in restoring some of the lost neurofilament motility. Additionally, similar to other neurotoxin treatments, the administration of certain antioxidants has shown some promise in reducing neurotoxicity of ingested arsenic.

Ammonia

Ammonia toxicity is often seen through two routes of administration, either through consumption or through endogenous ailments such as liver failure. One notable case in which ammonia toxicity is common is in response to cirrhosis of the liver which results in hepatic encephalopathy, and can result in cerebral edema (Haussinger 2006). This cerebral edema can be the result of nervous cell remodeling. As a consequence of increased concentrations, ammonia activity in-vivo has been shown to induce swelling of astrocytes in the brain through increased production of cGMP (Cyclic Guanosine Monophosphate) within the cells which leads to Protein Kinase G-mediated(PKG) cytoskeletal modifications. The resultant effect of this toxicity can be reduced brain energy metabolism and function. Importantly, the toxic effects of ammonia on astrocyte remodling can be reduced through administration of L-carnitine. This astrocyte remodeling appears to be mediated through ammonia-induced mitochondrial permeability transition. This mitochondrial transition is a direct result of glutamine activity a compound which forms from ammonia in-vivo. Administration of antioxidants or glutaminase inhibitor can reduce this mitochondrial transition, and potentially also astrocyte remodeling.

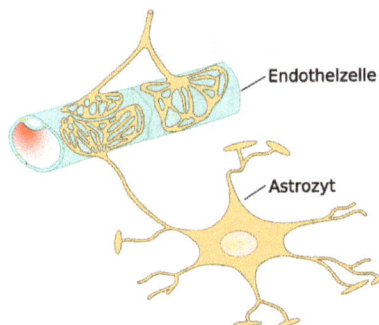

An Astrocyte, a cell notable for maintaining the blood brain barrier

Calcium-mediated Cytotoxicity

Lead pipes are common sources of ingested lead.

Lead

Lead is a potent neurotoxin whose toxicity has been recognized for at least thousands of years. Though neurotoxic effects for lead are found in both adults and young children, the developing brain is particularly susceptible to lead-induced harm, effects which can include apoptosis and excitotoxicity. An underlying mechanism by which lead is able to cause harm is its ability to be transported by calcium ATPase pumps across the BBB, allowing for direct contact with the fragile cells within the central nervous system. Neurotoxicity results from lead's ability to act in a similar manner to calcium ions, as concentrated lead will lead to cellular uptake of calcium which disrupts cellular homeostasis and induces apoptosis. It is this intracellular calcium increase that activates protein kinase C (PKC), which manifests as learning deficits in children as a result of early lead exposure. In addition to inducing apoptosis, lead inhibits interneuron signaling through the disruption of calcium-mediated neurotransmitter release.

Neurotoxins With Multiple Effects

Ethanol

As a neurotoxin, ethanol has been shown to induce nervous system damage and affect the body in a variety of ways. Among the known effects of ethanol exposure are both transient and lasting consequences. Some of the lasting effects include long-term reduced neurogenesis in the hippocampus, widespread brain atrophy, and induced inflammation in the brain. Of note, chronic ethanol ingestion has additionally been shown to induce reorganization of cellular membrane constituents, leading to a lipid bilayer marked by increased membrane concentrations of cholesterol and saturated fat. This is important as neurotransmitter transport can be impaired through vesicular transport inhibition, resulting in diminished neural network function. One significant example of reduced inter-neuron communication is the ability for ethanol to inhibit NMDA receptors in the hippocampus, resulting in reduced long-term potentiation (LTP) and memory acquisition. NMDA has been shown to play an important role in LTP and consequently memory formation. With chronic ethanol intake, however, the susceptibility of these NMDA receptors to induce LTP increases in the mesolimbic dopamine neurons in an inositol 1,4,5-triphosphate (IP3) dependent manner. This reorganization may lead to neuronal cytotoxicity both through hyperactivation of postsynaptic neurons and through induced addiction to continuous ethanol consumption. It has, additionally, been shown that ethanol directly reduces intracellular calcium ion accumulation through inhibited NMDA receptor activity, and thus reduces the capacity for the occurrence of LTP.

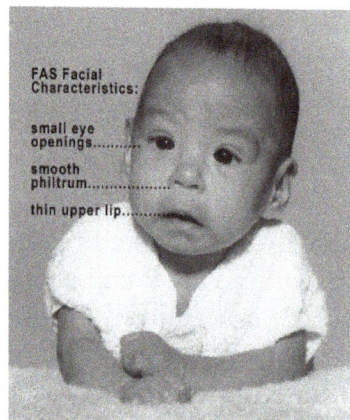

In addition to the neurotoxic effects of ethanol in mature organisms, chronic ingestion is capable of inducing severe developmental defects. Evidence was first shown in 1973 of a connection between chronic ethanol intake by mothers and defects in their offspring. This work was responsible for creating the classification of fetal alcohol syndrome; a disease characterized by common morphogenesis aberrations such as defects in craniofacial formation, limb development, and cardiovascular formation. The magnitude of ethanol neurotoxicity in fetuses leading to fetal alcohol syndrome has been shown to be dependent on antioxidant levels in the brain such as vitamin E. As the fetal brain is relatively fragile and susceptible to induced stresses, severe deleterious effects of alcohol exposure can be seen in important areas such as the hippocampus and cerebellum. The severity of these effects is directly dependent upon the amount and frequency of ethanol consumption by the mother, and the stage in development of the fetus. It is known that ethanol exposure results in reduced antioxidant levels, mitochondrial dysfunction (Chu 2007), and subsequent neuronal death, seemingly as a result of increased generation of reactive oxidative species (ROS). This is a plausible mechanism, as there is a reduced presence in the fetal brain of antioxidant enzymes such as catalase and peroxidase. In support of this mechanism, administration of high levels of dietary vitamin E results in reduced or eliminated ethanol-induced neurotoxic effects in fetuses.

n-Hexane

n-Hexane is a neurotoxin which has been responsible for the poisoning of several workers in Chinese electronics factories in recent years.

Receptor-selective Neurotoxins

MPP+

MPP+, the toxic metabolite of MPTP is a selective neurotoxin which interferes with oxidative phosphorylation in mitochondria by inhibiting complex I, leading to the depletion of ATP and subsequent cell death. This occurs almost exclusively in dopaminergic neurons of the substantia nigra, resulting in the presentation of permanent parkinsonism in exposed subjects 2–3 days after administration.

Endogenous Neurotoxin Sources

Unlike most common sources of neurotoxins which are acquired by the body through ingestion, endogenous neurotoxins both originate from and exert their effects in-vivo. Additionally, though most venoms and exogenous neurotoxins will rarely possess useful in-vivo capabilities, endogenous neurotoxins are commonly used by the body in useful and healthy ways, such as nitric oxide which is used in cell communication. It is often only when these endogenous compounds become highly concentrated that they lead to dangerous effects.

Nitric Oxide

Though nitric oxide (NO) is commonly used by the nervous system in inter-neuron communication and signaling, it can be active in mechanisms leading to ischemia in the cerebrum (Iadecola 1998). The neurotoxicity of NO is based on its importance in glutamate excitotoxicity, as NO is generated in a calcium-dependent manner in response to glutamate mediated NMDA activation, which occurs at an elevated rate in glutamate excitotoxicity. Though NO facilitates increased blood flow to potentially ischemic regions of the brain, it is also capable of increasing oxidative stress, inducing DNA damage and apoptosis. Thus an increased presence of NO in an ischemic area of the CNS can produce significantly toxic effects.

Glutamate

Glutamate, like nitric oxide, is an endogenously produced compound used by neurons to perform normally, being present in small concentrations throughout the gray matter of the CNS. One of the most notable uses of endogenous glutamate is its functionality as an excitatory neurotransmitter. When concentrated, however, glutamate becomes toxic to surrounding neurons. This toxicity can be both a result of direct lethality of glutamate on neurons and a result of induced calcium flux into neurons leading to swelling and necrosis. Support has been shown for these mechanisms playing significant roles in diseases and complications such as Huntington's disease, epilepsy, and stroke.

Dopamine

Dopamine is an endogenous compound that is used as a neurotransmitter to modulate reward expectation. Dopamine kills dopamine-producing neurons by interfering with the electron transport chain in neurons. This interference results in an inhibition of cellular respiration, leading to neuron death..

References

- Polychlorinated biphenyls and terphenyls, Environmental Health Criteria monograph No. 002, Geneva: World Health Organization, 1976, ISBN 92-4-154062-1

- Flame retardants: a general introduction, Environmental Health Criteria monograph No. 192, Geneva: World Health Organization, 1997, ISBN 92-4-157192-6

- Gore AC. Endocrine-Disrupting Chemicals: From Basic Research to Clinical Practice (Contemporary Endocrinology). Totowa, NJ: Humana Press; 2007. (Contemporary Endocrinology). ISBN 1-58829-830-2.

- "Most Plastic Products Release Estrogenic Chemicals: A Potential Health Problem That Can Be Solved". Environmental Health Perspectives. 2011-03-02. Retrieved 2015-04-06.

- "FDA to Ban BPA from Baby Bottles; Plan Falls Short of Needed Protections: Scientists". Common Dreams. 2012-07-17. Retrieved 2015-04-06.

- "Effects of human exposure to hormone-disrupting chemicals examined in landmark United Nations report". Science Daily. 2013-02-19. Retrieved 2015-04-06.

- Epidemiology and Statistics Unit (July 2011). "Trends in Tobacco Use" (PDF). American Lung Association. Retrieved 2015-04-02.

- Ing-Marie Olsson (24 November 2014). "The Cost of Inaction: A Socioeconomic analysis of costs linked to effects of endocrine disrupting substances on male reproductive health". Retrieved 10 October 2015.

- "Cotton Products Research: Durable Press and Flame Retardant Cotton". National Historic Chemical Landmarks. American Chemical Society. Retrieved 2014-02-21.

- "Bisphenol A (BPA): Use in Food Contact Application". News & Events. United States Food & Drug Administration. 2012-03-30. Retrieved 2012-04-14.

- ScienceDaily. 99% of pregnant women in US test positive for multiple chemicals including banned ones, study suggests; 14 January 2011 [Retrieved 1 February 2012].

- Brown, Eryn. "Jury still out on BPA, World Health Organization says", Los Angeles Times, 11 November 2010. Retrieved 7 February 2011.

Permissions

Index